THE GOOGLE STORY

Also by David A. Vise

Eagle on the Street
(with Steve Coll)

The Bureau and the Mole

Sweet Redemption

US: Google forms alliance with China firm for AdWords biz - Forbes - 10 hou[illegible]

oogle
ables users to search the Web, Usenet, and images. Features include PageRank,
d translation of results, and an option to find similar pages.
w.google.com/ - 3k - Aug 7, 2005 - Cached - Similar pages

www.google.com/search
Similar pages
[More results from www.google.com]

THE Google STORY

oogle News
gregated headlines and [illegible] search engine of [illegible] news sources.
w.google.com/ - 101k - Aug 7, 2[illegible]

oogle Groups
arn more about Google Groups. Create, search, [illegible] browse groups to discuss an
as. ... ©2005 Google - Searching [illegible] messages.
oups.google.com/ - 14k - Aug 7, 2005 - Cached - Similar pages

oogle Maps
ovides directions, interactive maps, and satellite/aerial imagery of the United States
arch by keyword such as type of business.
ps.google.com/ - 19k - A[illegible] Similar pages

DAVID A. VISE

AND

MARK MALSEED

oogle Scholar
ogle Scholar. Advanced Scholar Search ... Scholar Help. Stand on the shoulders
ogle Home - About Google - About Google Scholar. ©2005 Google.
holar.google.com/ - 3k - Cached - Similar pages

oogle Weblog
w: Google Earth. Formerly Keyhole, now free. Fly around the world from your ... Th
eo plugin is now available for free for Windows 2000 or ...
ogle.blogspace.com/ - 8k - Aug 7, 2005 - Cached - Similar pages

DELACORTE PRESS

oogle
ogle ... Advertising Programs - Business Solutions - About Google - Go to Google
005 Google - Searching [illegible] web pages
w.google.ca/ - 3k - Aug 7, 2005 - Cached - Similar pages

oogle
e local version of this pre-eminent search engine, offering UK-specific pages as we
sults.
w.google.co.uk/ - 3k - Aug 7, 2005 - Cached - Similar pages

oogle Toolbar
egrates with Internet Explorer's toolbar. Features include web search, image search

THE GOOGLE STORY
A Delacorte Book / November 2005

Published by
Bantam Dell
A Division of Random House, Inc.
New York, New York

Poem on page 27 and GLAT in Appendix II used courtesy of Google.

Book design by Glen Edelstein

Library of Congress Cataloging in Publication Data is on file with the publisher

ISBN-10: 0-553-80457-X
ISBN-13: 978-0-553-80457-7

Printed in the United States of America
Published simultaneously in Canada

www.bantamdell.com

10 9 8 7 6 5 4 3 2 1
BVG

To Lori—
My search was over the day I met you

To my parents, Roger and Zoriana

Contents

THE GOOGLE STORY

Introduction

Not since Gutenberg invented the modern printing press more than 500 years ago, making books and scientific tomes affordable and widely available to the masses, has any new invention empowered individuals, and transformed access to information, as profoundly as Google. With its colorful, childlike logo set against a background of pure white, Google's magical ability to produce speedy, relevant responses to queries hundreds of millions of times daily has changed the way people find information and stay abreast of the news. Woven into the fabric of daily life, Google has seemingly overnight become indispensable. Millions of people use it daily in more than 100 languages and have come to regard Google and the Internet as one. The quest for immediate information on anything and everything is satisfied by "googling" it on a computer or cell phone. Men, women, and children have come to rely so heavily on Google that they cannot imagine how they ever lived without it.

Google's transcendent and seemingly human qualities give it special appeal to an amazingly wide range of computer users, from experts to novices, who trust the brand that has become an extension of their brains. That appeal is universal, enabling it to overcome differences in culture, language, and geography en route to becoming a global favorite. For a young firm that has not spent money to advertise or promote its brand name, these are unparalleled achievements. Google's growth has occurred entirely by word of mouth, as satisfied users recommend it to their friends, and others learn about it through the media and online. No Madison Avenue marketers have pushed it. Instead, people have come to feel emotionally attached to the search engine, calling on it whenever they wish to satisfy their interest or curiosity. In an uncertain world, Google reliably provides free information for everyone who seeks it. It is a seductive form of instant gratification for their minds.

Most Google users have no idea how the search engine was created, what makes it so profitable and valuable, why it has triumphed over deeper-pocketed competitors, and where it is heading in the future. In the pages of this book, we will answer all these questions for the first time. Until now, most of the answers have remained secret, hidden deep inside the Googleplex, the company's space-age Silicon Valley campus.

John Hennessy, a top computer scientist who is president of Stanford University and a Google board member, says the firm is unique in today's bifurcated world of sophisticated software and hardware companies because it is a leader in both areas. To power its search and search-related services, Google runs patented, custom-designed programs on hundreds of thousands of machines that it also custom builds. The optimal blending of technologies by the world's most innovative company produces superior search results instantaneously. No word in the English language exists to describe this seamless melding of hardware and software at such a massive scale, so we have named it Googleware.

Hennessy says that the most important technological advantage distinguishing Google from would-be competitors is that its employees assemble and customize all of the personal computers the company uses to carry out searches. This is perhaps Google's best-kept secret. Experts generally regard personal computers as commodity products, akin to toasters, but Google assembles, deploys, and is constantly improving the performance capabilities of more than 100,000 inexpensive PCs. It builds and stacks them atop one another in refrigerator-size racks, stringing them together with patented software and wiring. No enterprise has more computing power than Google with its network of garden-variety PCs on steroids.

"They run the largest computer system in the world," Hennessy said. "I don't think there is even anything close."

In an age of specialization of labor, Google secretly assembles each and every PC in its massive network inside secure facilities that are strictly off-limits to outsiders, including visitors to the Googleplex who think they have seen it all. Google is able to do this affordably because the massive scale of its operation makes it cost-effective and of higher quality than buying custom PCs from someone else. Working together, these customized computers rapidly carry out searches by breaking the queries down into tiny parts. These parts are processed simultaneously by comparing them to copies of the Internet that have been indexed and organized in advance.

With plenty of redundancy built into its network, Google is able to reliably fire up more and more computers daily, rapidly returning search results without a glitch—and without human intervention—even as some PCs burn out and are not replaced. Instead, other PCs take over. Remarkably, as Hennessy noted, there is no comparable computer network or database in the public or private sector anywhere in the world.

"They realized early on that if they did a good job with the hardware, that it could be a competitive strength for the company," he said. "The hardware would be the key factor in being able to do search efficiently. And they are replicating computers at such a large volume that it makes sense to do it themselves. It is why launching a search engine that is competitive with Google would require a much larger capital investment than people realize."

Thanks to the inclusion of small, highly targeted text advertisements that searchers click on for information, Google the search engine became Google the money machine a few years ago. This was while Google was still private, and thus inscrutable to all but the innermost circle of early investors and employees. But once its stunningly rapid growth and billions of dollars in ad-generated profits became public knowledge, a growing legion of investors recognized that something unique was going on and sought to own a part of it. On August 19, 2004, Google went public in an unconventional initial public offering at $85 per share, raising nearly $2 billion in the largest technology IPO ever. In less than a year, the stock soared to more than $300 per share, making it a financial and technology powerhouse without peer. In the seven years after Google's founding in 1998, Microsoft stock did not increase in value at all. During that same period, Google shares came to be worth more than $80 billion.

If by chance you missed out on this bonanza, remember that the experts did too. Blue-chip venture capital firms, Yahoo!, AltaVista, and many other major search engine and technology companies approached by Stanford University turned down the chance to buy Google's search system for $1 million. Their refusals forced Stanford Ph.D. students Sergey Brin and Larry Page to reluctantly drop out of school and start the firm. By the summer of 2005, each of the founders had a net worth of more than $10 billion.

Hennessy recalled the first time a Stanford professor brought Google to his attention in the mid-1990s. Like Brin and Page, Hennessy had grown frustrated using AltaVista, the best search engine at the time. While it did a reasonably good job of canvassing the Internet, it did a poor job of ranking search results. Hennessy remembers hearing that the creators of a search engine on campus, through a new mathematical formula known as PageRank, had come up with a way to give users the most important search results in a flash.

As a computer scientist of note, he typed his own name into the Google search box to see what would happen. "The first thing that came up for me was Stanford University," Hennessy said. "That didn't happen on the other search engines."

In the pages of this book, we will illuminate the business strategies and the corporate motto, "Don't Be Evil," that Google employed on the way to establishing a cadre of advertisers and Web sites with strong ties to the search engine. These Google affiliates, numbering in the hundreds of thousands, have enormous vested financial interest in the search engine's continuing growth and success. We have named this robust and self-reinforcing network "the Google Economy," and it has propelled the search engine to grow seven-fold within a year of its going public. As long as Internet use and online ad spending continue to meet projections for growth, the Google Economy is likely to continue expanding.

Google itself appears destined for growth and expansion, given the relative infancy of the Internet as a mass medium, and the growing migration of advertising dollars from television and print to the online world. The reason is simple. Google figured out how to make advertising on the Internet work effectively by targeting it to individuals at the moment they most need it: when searching for information.

Google's success depends on the continuing day-to-day involvement of cofounders Brin and Page. Googleware and the lucrative Google ad system are a reflection of their genius and foresight. Going forward, it is the founders' focus, leadership, and grand ambition that are the most important ingredients in Google's long-term success. With them at the helm, Google is likely to remain a popular search engine that people connect to both electronically and emotionally. Nevertheless, suspicion of Google will increase as it grows in size and records more of our online behavior, and there will be calls, eventually, for federal regulation. Another problem is that Google profits from fraudulent clicks on its text ads, a challenge it has yet to bring under control.

Together, Brin and Page power Google, breathing life into its interactions with hundreds of millions of users daily. They constantly motivate the collection of brilliant mathematicians, engineers, and technologists at the Googleplex to tackle larger and larger problems. And they push hard to give computer users a crack at using these products before they are perfected, a method that keeps innovation humming while providing valuable feedback to get rid of flaws.

Google is free to users. The billions of dollars in money and profits that flow from it are a by-product of the company's concentrated efforts at innovation, rather than a yardstick used internally to measure success or to determine whether a project is worthy of exploration. Unlike most companies, where executives and product managers try to think of ways to make money and then create products, Google is a place where technologists think first of ways to solve problems; only later, if ever, do they worry about how to "monetize" them. Dedicated teams of engineers are encouraged to dream up entirely new ideas to make the search engine operate faster and better. One reason the company has no need for marketing is that its culture fosters a laserlike focus on serving the best interests of Google users. They, in turn, become its best advocates.

Google does not seek to make as much money as it could in the short run. The most obvious example of this is the Google homepage, considered the most valuable piece of real estate on the Internet. A quick trip to www.google.com will confirm that Google displays no advertising on this page, forgoing tens of millions of dollars in revenue and profits to give users a higher-quality search experience.

The soul of the Google machine is rapid innovation, the most important subject discussed at nearly every board meeting of the firm. To Brin and Page, sustaining innovation as Google grows is their foremost challenge, for innovation is the reason the company raced ahead of others and stays out in front. Its founders are keenly aware that someone, somewhere, is always attempting to find a better, faster, and smarter way to do things. And maintaining smart innovation amid torrid growth is a complex undertaking that has vexed other young enterprises of enormous promise.

When it comes to products, Brin and Page operate in a hands-on manner, pushing hard for the introduction of new features and offerings at a blistering pace. Some of these changes are highly visible; others, hidden from view, provide users with better and speedier search results. The founders have time to focus heavily on what matters to users because their colleague, chief executive officer Eric Schmidt, tends to the company's business affairs. This sets them free to do what they do best. However, the entire trio at the top gets involved in both business affairs and product development. Brin, it turns out, is quite a good deal-maker and Page is adept at finding ways to cut costs, notably in the vitally important area of electricity needed to power and cool Google's computers.

At Google, the preference is for working in small teams of three, with individual employees expected to allot 20 percent of their time to exploring whatever ideas interest them most. The notion of "20 percent time" is borrowed from the academic world, where professors are given one day a week to pursue private interests. Because the company lacks the usual layers of middle management, the hierarchical structure found in traditional corporations is nonexistent. At the same time, no other company has Google's combination of brains, immense computing resources, and focus on long-term results, vaulting it into a league of its own. These days, other companies have a tough time competing against Google for employees.

Operating at the epicenter of the Internet, Google is the place with the buzz where the best and brightest engineers on the planet flock to work. They are leaving universities, NASA, Bell Labs, Microsoft, and elsewhere for a dynamic setting that more nearly resembles a graduate school campus than a traditional business headquarters. And the company's skyrocketing stock price since its initial public offering has captured worldwide attention, assisting in the recruiting effort.

The enormous media attention focused on Google is a function of the search engine's popular appeal, its place in the zeitgeist of the era, and the heavy use reporters and editors make of it in their day-to-day work. To *google* means "to search." That the company's name has become a verb in English, German, and other languages is testament to its pervasive influence on global culture.

Google's potential appears to be greater than that of iconic companies that have preceded it. Over the decades, a series of technologies have swept across the landscape, each wave larger than the one that came before. IBM and mainframes solved the data-processing problem for corporations decades ago. Then came Intel and Microsoft, both of which made enormous contributions to the personal computer and gave individuals a new source of power, ultimately catapulting the PC industry to greater penetration and profitability than the mainframe industry. Now the Internet, originally a Defense Department project, has emerged as the platform of choice, vaulting Amazon, Yahoo, eBay, and Google to the forefront. Among the icons distinctive to this wave, none is riding higher than Google, the only new megabrand created among Internet firms in the past decade. The company's stock is a bellwether of investor confidence in the future of the Internet and the particular business model Google has created to capture targeted advertising dollars.

Two of the most compelling areas that Google and its founders are quietly working on are the promising fields of molecular biology and genetics. Millions of genes in combination with massive amounts of biological and scientific data are an excellent match for the Google search engine, the tremendous database the company has in place, and its immense computing power. Already, Google has downloaded a map of the human genome and is working closely with biologist Dr. Craig Venter and other leaders in genetics on scientific projects that may lead to important breakthroughs in science, medicine, and health. In other words, we may be heading toward a time when people can google their own genes.

"This is the ultimate intersection of technology and health that will empower millions of individuals," Venter said. "They have the most computing power of any enterprise. The scale they deal with is far beyond what the government databases are now. Helping people understand their own genetic code and statistical code is something that should be broadly available through a service like Google within a decade."

According to Brin and Page, Google is not a conventional company and does not intend to become one. If Google were a person, said Brin, it would have started elementary school around August

19, 2004, the date the company went public, and it would have just finished the first grade in the summer of 2005. Google is in the process of digitizing millions of books from the libraries at Stanford, Harvard, the University of Michigan, the New York Public Library, and Oxford. Its goal is to put as many of these books as possible online and make them searchable, tearing down the physical limitations of libraries. By itself, this is an exceedingly ambitious undertaking, and it has positive and far-reaching educational and social consequences globally.

Google and its founders have traveled far. At the same time, both they and the company are still young—and this is just the beginning.

CHAPTER 1

A Healthy Disregard for the Impossible

Sergey Brin and Larry Page cruised onto the stage to the kind of roars and excitement that teenagers normally reserve for rock stars. They had entered the auditorium through a rear door, leaving behind photographers, sunglasses, a pair of hired cars with drivers, and an attractive young woman who was traveling with Sergey. Dressed casually, they sat down and cracked smiles, pleased at their heroes' welcome. They were near the birthplace of civilization, thousands of miles and an ocean away from the place where their work together had begun. It seemed as good a place as any for a pair of young superstars, whose shared ambition revolved around changing the world, to talk about what they had done, how they had done it, and what their dreams were for the future.

"Do you guys know the story of Google?" Page asked. "Do you want me to tell it?"

"Yes!" the crowd shouted.

It was September 2003, and the hundreds of students and faculty at this Israeli high school geared toward the brightest young minds in mathematics wanted to hear everything the youthful inventors had to say. Many of them identified with Brin because, like him, they had escaped with their families from Mother Russia in search of freedom. And they related to Page just as eagerly, since he was part of the duo that had created the most powerful and accessible information tool of their time—a tool sparking change that was already sweeping the world. Like kids playing basketball and dreaming of being the next Michael Jordan, the students wanted to be like Sergey Brin and Larry Page.

"Google was started when Sergey and I were Ph.D. students at Stanford University in computer science," Page began, "and we didn't know exactly what we wanted to do. I got this crazy idea that I was going to download the entire Web onto my computer. I told my advisor that it would only take a week. After about a year or so, I had some portion of it." The students laughed.

"So optimism is important," he went on. "You have to be a little silly about the goals you are going to set. There is a phrase I learned in college called, 'Having a healthy disregard for the impossible,'" Page said. "That is a really good phrase. You should try to do things that most people would not."

As proponents of tackling important problems and seeking transformative solutions, Brin and Page were certainly armed with a healthy disregard for the impossible. And while not much older than the throng of high school students who packed the jammed auditorium, they were truly in a class by themselves. In the rich and storied history of American invention and capitalism, there had never been a meteoric rise comparable to theirs. It had taken Thomas Edison a quarter century to invent the lightbulb; Alexander Graham Bell had spent many years developing the telephone; Henry Ford created the modern assembly line and turned it into the mass production and consumption of automobiles only after decades of work; and Thomas Watson Jr. labored long and hard before IBM rolled out the modern computer. But Brin and Page, in just five years, had taken a graduate school research project and turned it into a multibillion-dollar enterprise with global reach. They were in Tel Aviv, but had it been Tokyo, Toronto, or Taipei, the Google Guys would have received the same raucous reception.

The youthful pair had changed the lives of millions of people by giving them free, instant access to information about any subject. And by being devilishly clever in the Internet age, they had created the best-known new brand in the world without advertising to promote the name. The two were astute businessmen, and knew that to succeed over time it was imperative that they remain in complete control of their privately owned business and its quirky culture. It saddened Page that many inventors die without ever seeing the fruits of their labors. Determined to avoid a similar fate, he and Brin understood how to use the right connections, access to money and brilliant minds, raw computing power, and a culture of limitless possibilities to make Google a beacon and a magnet. In the click of a mouse, it had replaced Microsoft as the place for the world's top technologists to work. Yet they knew that maintaining the pace of innovation and the mantle of leadership would be no easy feat, since they faced a deeper-pocketed competitor in Microsoft, and a ruthless combatant in its chief, the billionaire Bill Gates.

Supremely confident about their achievements and vision, Brin and Page had been on a roll ever since they started working together. They wanted no one—neither competitors nor outside investors—to come between them or interfere in any way. That combination of dependence on each other, and independence from everyone else, had contributed immeasurably to their astounding success.

"So I started downloading the Web, and Sergey started helping me because he was interested in data mining and making sense of the information," Page went on, continuing with the pair's history. "When we first met each other, we thought the other was really obnoxious. Then we hit it off and became really good friends. That was about eight years ago. And we started working really, really hard on it." He stressed this critical point: inspiration still required plenty of perspiration. "There was an important lesson for us. We worked through holidays, and worked many, many hours a day. It ended up working out, but it is hard because it takes a lot of effort."

Page said that as they told friends about Google, more and more people started using it. "Pretty soon we had 10,000 searches a day at Stanford. And we sat around the room and looked at the machines and said, 'This is about how many searches we can do, and we need more computers.' Our whole history has been like that. We always need more computers."

It was a sentiment the students and teachers at the school could relate to.

"So, we started a company. Being in Silicon Valley at this time, it was relatively easy to do. You have a number of very excellent companies here as well," he added, alluding to the growing technology sector in Israel, "so this is also a good environment to do things like that. We started up the company, and it grew and grew and grew—and that is why we are here. So that," he concluded, "is 'The Google Story.'"

But there was something more he wanted to convey: closing words of inspiration.

"Let me explain to you guys why I am so excited about being here," Page said. "And it is really that there is so much leverage in science and technology. I think most people don't really realize that. There is so much that can be done with these new technologies. We are an example of that.

"Two kind of crazy kids have had a big impact on the world because of the power of the Internet, the power of the distribution, and the power of software and computers. And there are so many things like that out there. There are so many opportunities where you can have a huge impact on the world by using the leverage of science and technology. All of you are uniquely positioned, and you should be excited about that."

Brin interjected that the pair's overseas travels included not only Israel but also several European countries. They were on the prowl for talent, and they were considering opening new offices. For Sergey, who has a sharp sense of humor, the search was ongoing.

"We spend most of our time trying to get Internet access," he quipped. "We surf every day. I was on until 4 A.M. last night. And then I got on again earlier this morning. It is an invaluable tool. It is kind of like a respirator now."

Having fled Russia with his family for freedom from anti-Semitism and discrimination, Brin had something powerful in common with the experience of many of the Russian-born students at the Israeli high school. His father, Michael Brin, had captured the essence of why he and his wife and young son had left Russia when he said that one's love of country is not always reciprocated. Sergey recognized that he had a special opportunity to motivate these students, so he took the wireless microphone in hand, stood up, and connected. He had been advised before they came to the high school that this group truly was extraordinary, the best of breed, and the recent recipients of all but three of the top mathematics prizes in the country.

"Ladies and gentlemen, girls and boys . . ." he said, before speaking in Russian, to the delight of the students, who broke out in spontaneous applause.

"I came and emigrated from Russia when I was six. I went to the United States. Similar to here, I have standard Russian-Jewish parents. My dad is a math professor. They have a certain attitude about studies. And I think I can relate that here, because I was told that your school recently got seven out of the top ten places in a math competition throughout all Israel."

The students, unaware of what was coming next, applauded their seven-out-of-ten achievement and the recognition from Sergey.

"What I have to say," Sergey continued, "is in the words of my father, 'What about the other three?'"

A ripple of laughter went through the crowd.

"You have several things here that I didn't have when I was going through high school. The first one is the beautiful weather and the windows. My school in Maryland, which was built during the '70s energy crisis, has three-foot-thick walls and no windows. You are very fortunate to be in such a beautiful setting. The other thing we didn't have back then was Internet access.

"Let me get a show of hands. How many of you used the Internet yesterday?"

Virtually every hand in the auditorium shot up.

"Anybody do any searches?"

Once again, the hands went up.

Brin knew that technology-savvy and information-hungry Israelis loved Google and used it heavily, but he couldn't resist having a little fun with the students. He has a playful side to his personality that many found endearing—but he has a mischievous side too, which some found baffling because they couldn't always tell when he was being serious, when he was being too clever for them to follow, and when he was just kidding around.

"Did you use AltaVista?" he asked, raising his right hand to signal students what to do if they had used the first major search engine that Google had left in the dust.

"Excite?" he asked, listing another erstwhile search engine.

"Just curious," Brin went on, without breaking cadence, "what search engine you used here."

Without a pause, which would have revealed that he was joking with the students, he returned to the message he had begun delivering.

"Growing up I didn't have the Internet, or not in its current form, and the World Wide Web," he said. "Today, the world is very different, because each of you has the power to get information about any subject in the world. And that is very, very different from when I went to school."

For a moment, imagine yourself a bright teenager, taking in every word delivered by Sergey Brin. You listen carefully to him not only because of what he already has done, but also because he is a youthful Russian. You think about where he seems to be going through a blend of moxie and technology, and what he is telling you about yourself and your own potential.

"You really have a lot of power that our generation did not," Brin said. "I think that will enable you to succeed earlier in life, and much more in life, than I did.

"Finally, the most important thing I didn't have growing up, which all of you do, is all of your wonderful peers who are also very hardworking and inspiring. I shouldn't say [bad] things about my peers. But my school was a little bit different. It was in the suburbs of Washington DC, and the academic standards and the children's goals were different. You should really value what you have here with such an incredibly talented peer group. I am sure you are going to know many of your classmates for the rest of your lives."

Brin and Page wrapped up their talk and indicated to the assembled students that it was time for questions. None were aware of the commotion that had just kicked up outside the auditorium, or whom it involved, or what it would soon mean for all of them.

Do you think Google was the highlight of your career? the first student asked Brin and Page.

"I think it was the smallest of accomplishments that we hope to make over the next 20 years," said Brin. "But I think that if Google is all we create, I don't think I would be very disappointed."

Page felt otherwise. "I would be very disappointed," he said. "We have many years left to go."

The next student asked about new projects under way at Google.

"We run Google a little bit like a university," Brin explained. "We have lots of projects, about 100 of them. We like to have small groups of people, three or so people, working on projects. Some of them, for example, are related to molecular biology. Others involve building hardware. So we do lots of stuff. The only way you are going to have success is to have lots of failures first."

The students applauded. The idea of eventually succeeding after failing first, of not being afraid to fall down, resonated with them immediately. Israelis are risk takers by nature, Russians much less so. Yet the families of these Russian students had risked the unknown by leaving their homes in search of a better life. Now here they were, sitting with two of the world's leading thinkers, probing and laughing in ways that reflected the freedoms they enjoyed and that seemed just as important to the Google founders.

Since the question had not really been answered, another student asked again about new Google projects.

"We are embarrassed about our new projects," Brin said, teeing up a joke—maybe. "One very recent one comes from Israel. Yossi

[Vardi, the inventor of instant messaging] has a friend who makes underwear, Calvin Klein underwear. So we are trying to see if we can forge a partnership to have Google-branded underwear.

"If Google has underwear, who will buy it?" he asked. Hands shot up. "We will have to send a shipment to the school," Brin said, adding, "That is probably one of the less technical projects we have."

"A project that is closer to our hearts," he continued, "involves translation, and I was asked about Hebrew to English, which is not one of the ones we are working on. One of the most important aspects of information access is that a lot of information is not in languages that you understand. We have an existing translation service, but we think we can make it exceptionally better."

What, another student asked, does the name Google mean?

"*Google* means a very large number," Page said. "It is the number '1' followed by 100 zeros. And we were trying to come up with a name. We were confused about how to spell this, and so we actually spelled it incorrectly. It is a mathematical term and it is spelled g-o-o-g-o-l. So that is the right way Google is spelled. Most people don't know that, fortunately."

Are there any plans, a student asked, to promote commercial sites in Google's search results?

"We don't let that be influenced by any business relationships," Brin said, adding that the search engine's results were "unbiased."

That prompted a new question: How does Google make money?

"Google gets paid for every search that happens, more or less, mostly through advertising," Page said. "People pay for advertising. We are very lucky in that we chose to make ads relevant rather than having flash banner ads. It helps us have the best search engine. We also get paid by other companies, like AOL, that use our search engine."

A student asked what they thought about pornography, saying it was all over the Internet and that people used Google to find it.

"Naturally, it depends on what you search for," Sergey replied, apparently dodging the question; then he continued. "I want to tell you about something else Larry and I share in common. Both of us at a very early age went to something called a Montessori school. The theory of Montessori is you let kids do a lot of what they want when they are six, seven, eight, nine, ten, eleven, twelve. But after that, because of the hormones that boys have, you actually need to send them to do hard labor in their teens. Otherwise, their mind gets distracted." Brin seemed to be subtly answering the pornography question after all. "In this case, there are many ways to apply yourselves. You don't have to do hard labor on the farm—but it is important to maintain focus, even through these difficult years."

Another student asked about Google's competition.

"It started off being Excite, AltaVista, and others," Page said. "They were not that focused on search, so we didn't have as much trouble with them as we could have. Nowadays, we get much larger competition and it is a bigger challenge for us. Google is going to probably be a medium-size technology company. We have over 1,000 people who work at Google. We are starting to have offices all around the world. That is part of why we are traveling around. This is the real test for us. The tough part is whether we can do it for the long-term, a 10-to-20-year-old company, or whether we will be overtaken."

"To just invent something and have a great idea is a lot of work," Page went on, "but it is not enough. You have to get it out in the world. At Google, it is a combination of scientific skills, and mathematical skills, and computer skills, and also very strong skills about how to get people excited about their work."

Just then, the eyes in the room darted away from Brin and Page to two men walking in the back door of the auditorium. A wave of excitement swept the room as the students, and Brin and Page too, realized who it was: Mikhail Gorbachev, the former prime minister of the Soviet Union, and Shimon Peres, the esteemed Israeli political leader who was celebrating his eightieth birthday that week. Gorbachev, whose policies had ignited the transformations that brought freedom and capitalism to Russia, was a hero to the gathered students for having helped make it possible for their families to seek new lives. The Russian students stood and cheered. Gorbachev, known for his dour expression, reached out to them, and to Brin and Page, with a look of satisfaction.

"I understand this school is like a mini–Soviet Union," he said, addressing the audience in Russian. "There are people here from all 15 republics. I'm glad you are friendly to one another and support one another. I wish you success in your education, and I'd like to say that today we have young guests from the United States of America. These people illustrate what modern knowledge really is. And what that means. They have good, bright minds—I will say 'unique' minds. They have done a lot in life, even at their age. So today they are leaders and movers in the development of new technology. They became interested in this school and maybe will help in some way to bring the educational process to a modern level to help develop fine teenagers—to change people's lives to allow them to accomplish more than we did."

A smiling Gorbachev shook hands with Sergey and Larry, who were beaming. Here they were, barely 30 years old, sharing the stage with two elder statesmen, world leaders assured of a place in history. For a moment, that school stage became the world's stage, the place where past, present, and future converged with unlimited potential.

CHAPTER 2
When Larry Met Sergey

When Larry met Sergey in the spring of 1995, they connected instantly. Despite their differences, there was no denying the chemistry between them; the energy was palpable. It was during a new student orientation at Stanford, and Sergey was showing Larry and other prospective graduate students around the sun-laden California campus and its environs. Suddenly, the pair started to argue about random issues. It seemed an odd moment for two people who barely knew each other to be debating, sparks flying, but in fact each was playing a favorite game.

Younger than Larry and the others, Sergey had been at Stanford for two years already. A math whiz, he completed his undergraduate degree at 19, aced all ten of the required doctoral exams at Stanford on his first try, and teamed easily with professors doing research. Confident, fit, and outspoken, Sergey especially enjoyed the gymnastics, swimming, and social life at Stanford, while also spending time on computer and math problems. Larry, a Midwesterner, felt uneasy about being one of the select few admitted to Stanford's elite and competitive Ph.D. program, and harbored serious doubts that it would work out. He hoped to make friends with other students during the campus visit.

"At first it was pretty scary. I kept complaining," he recalled, "that I was going to get sent home on the bus."

Fortunately, Larry and Sergey quickly found each other, as well as a shared passion for jousting with an intellectually worthy adversary, even if it meant taking absurd positions. The subjects of their animated exchanges didn't matter. What counted was persuading the other guy to see the world your way.

Their relentless banter and verbal sparring laid the groundwork for what was to become a partnership imbued with mutual respect, even though each found the other cocky and obnoxious at first. Both had grown up in families where intellectual combat was part of the daily diet, especially, though not only, when it came to issues of computers, mathematics, and the future. Learning to vigorously defend compelling ideas had left each of them with an intellectual depth that belied their youth. Most people who got to know Larry and Sergey considered them bright, friendly, and somewhat goofy guys.

They had overlapping ambitions and interests, and complementary personalities and skills. Sergey, who had a much younger brother back home in the suburbs of Washington DC, was louder, an extrovert accustomed to being in the limelight. Larry, the younger of two brothers, was quieter and more contemplative. But once Larry and Sergey were back at Stanford for the 1995–1996 academic year, their intellectual dueling evolved into a lasting friendship. That transformation, it turned out, had much to do with all that had taken place long before they ever met.

In a world defined more by genetics and technology than by geography, Sergey Brin and Larry Page had something significant in common: they were both second-generation computer. They grew up using computers in their elementary school years, under the tutelage of parents who used computers and sophisticated mathematics at home and at work. This set Larry and Sergey apart from others their age. They also shared other commonalities: they attended Montessori schools, which accelerated their early education; they lived near major universities where their fathers were esteemed professors; and they had mothers whose jobs revolved around computing and technology. Scholarship was not just emphasized in their homes; it was treasured. That Larry and Sergey would go on to pursue postgraduate studies and then enter the academic and scientific communities, just as their parents had done, seemed inevitable.

Back in 1960, Larry's father, Carl Victor Page, had been one of the first undergraduates to receive a computer science degree from the University of Michigan. Five years later, he earned his Ph.D. in the novel field. Larry's mother, Gloria Page, was a database consultant with a master's degree in computer science. Larry admired his fun-loving, gregarious dad, who among other things took him to Grateful Dead concerts while he was growing up. Some 15 years before Larry went west, Carl Page spent time at Stanford as a visiting scholar. But he spent most of his career teaching at Michigan State University, where his wife also taught computer programming.

During Larry's second semester at Stanford, his father, a survivor of childhood polio, died at the age of 58 from complications of pneumonia. His health had deteriorated rapidly in a period of weeks before he stopped breathing and passed away in the hospital. For Larry, the sudden loss was traumatic. "I remember Larry sitting on the steps of the Gates Building, and he was very depressed," said Sean Anderson, an office-mate of Larry's at Stanford. "A number of his friends were around trying to comfort him."

Carl Page was eulogized as "a pioneer and world authority in computer science and artificial intelligence" and "a prolific scholar and beloved teacher and mentor to innumerable students." One of his colleagues at Michigan State, George Stockman, said Carl had been "looking forward to Larry becoming a professor." He also suggested that Larry's penchant for arguing was homegrown. "In some sense he was a little hard to deal with," Stockman said of his late colleague, "because he wanted to argue about everything and did, and . . . [he] shared a lot of that with his son. So intellectually they shared in a lot of discussion." That aspect of the father-son relationship offered a clue to the dynamics of Larry's early exchanges with Sergey.

Despite his grief, Larry remained enrolled at Stanford. It helped that his older brother, Carl Jr., lived and worked in Silicon Valley. They had each other, so Larry wasn't left to bear the loss alone, and the two spent time together, fondly recalling their dad and reflecting on their childhood memories.

This was not the first time Larry had been forced to deal with a painful sense of loss. Born on March 26, 1973, he lived through the divorce of his parents when he was eight years old, and the split affected him. His parents remained committed to raising Larry in as healthy a manner as possible. Ultimately, Larry came to feel that he was showered with love and wisdom from two mothers: his real mom, and Joyce Wildenthal, a Michigan State professor who had a long-term relationship with his dad. The two women remain on friendly terms to this day.

Larry attended East Lansing's MacDonald Middle School, where his Cub Scout leader, Nick Archer, remembers him as "an independent thinker." Carl Jr. recalls Larry as an inquisitive younger brother with wide-ranging interests that went well beyond the realm of computing. When they were growing up, both brothers were keenly interested in finding out "how things work," Carl Jr. recalled, "not just technical things, but social things, government, politics, everything." Their father had been a staunch, lifelong Democrat, preaching the party's gospel of education and opportunity for all. Their paternal grandfather had participated in the 1937–1938 auto workers' sit-down strike in Flint, Michigan, that empowered the labor movement, one of many factors that influenced their progressive approach to social issues. Their other grandfather had moved to Israel, becoming an early settler in the desert town of Arad, where water and other resources were scarce. He endured hardship, working in difficult surroundings as a tool and die maker near the Dead Sea. Long after his death in Israel, his pioneering spirit lives on in his grandsons.

While Larry's mother is Jewish, his father's religion was technology, and Larry was raised without any Jewish identity or religion. He took after his dad, who regularly exposed his sons to computers as part of their upbringing. "I was really lucky that my father was a computer science professor, which is unusual for someone my age," Larry said. "We were lucky enough to get our first home computer in 1978. It was huge, and it cost a lot of money, and we couldn't afford to eat well after that. I always liked computers because I thought you could do a lot with them."

In fact, he had so much exposure to computer technology at home that his schoolteachers in Lansing alternated between being amazed and baffled by the way he did homework. "I turned in a word processed document in my elementary school, and they didn't know what a dot matrix printer was. They were very confused about things," he said. Carl Jr. recalls Larry's proficiency when he was in only first grade: "One of the early things I remember Larry doing was typing *Frog and Toad Together* into his computer, one word at a time, when he was six years old." Several years later, Larry took a screwdriver set and dismantled the family's power tools. He also enjoyed helping Carl Jr.—who was nine years older—with his college computer homework when Carl came home from the University of Michigan during breaks. Before graduating from East Lansing High School, Larry also built a working inkjet printer out of Legos. "I never got pushed into it. I just really liked computers," he said.

Larry followed in the footsteps of his father and brother by attending the University of Michigan in Ann Arbor, where he studied computer engineering and took business courses en route to receiving his undergraduate degree in 1995. He served as president of the Michigan chapter of Eta Kappa Nu, the national honor society for computer engineering students, and had fun working in the group's campus doughnut stand.

He also took advantage of undergraduate leadership development programs. "In particular, the LeaderShape program was an amazing experience," he said. This university-wide program was aimed at giving Michigan students the skills they would need to be leaders in society.

Larry also said he learned much from superb professors. "I had access to amazing people who were willing to share their advice and expertise, and to help me succeed," he said. Those feelings were mutual. His Michigan professors viewed him as an excellent student. "Larry just stood out; he was always ahead," said Elliot Soloway, a professor of electrical engineering and computer science. "Larry used a handheld computer for his project in my course, before anyone knew what a handheld computer even was."

Sergey Brin's parents also have backgrounds rich in science and technology. His mother, Eugenia Brin, is an accomplished scientist at NASA's Goddard Space Flight Center. Among other things, she works on simulations related to atmospheric and weather conditions affecting space travel. His father, Michael Brin, teaches math at the University of Maryland and has published numerous academic papers exploring complex mathematical subjects ranging from abstract geometry to dynamic systems.

"He was a normal kid," Michael Brin said of Sergey, "but he wanted to be around computers, starting with games and the old Commodore 64s," an early personal computer.

Born in Moscow on August 21, 1973, young Sergey Brin left the Soviet Union with his parents when he was six. The family, fleeing anti-Semitism, was searching for greater freedom and opportunity for themselves and their son.

"I left because of myself and because of his future," said Michael Brin.

Sergey had at least one family member who had been in America many years before their arrival. In a move that broke the mold for women of her day, Sergey's great-grandmother pursued the study of microbiology at the University of Chicago. But in 1921 she joined others who believed in the communist ideal by returning to Moscow to assist with the building of a new Soviet state. Sergey's grandfather was a math professor in Moscow, as was his father, who received his Ph.D. in mathematics there. His mother, also a mathematician, worked in Russia as a civil engineer.

For ten years Michael Brin worked as an economist for

Gosplan, the Soviet central planning agency, where he was forced to produce propaganda with statistics to prove that the quality of life in the Soviet Union was better than in America.

"Most of that time I devoted to proving that Russian living standards were much, much higher than the American living standards," Michael Brin recalled. "I know a lot about numbers. I wasn't very precise."

After arriving in the U.S., Michael Brin began teaching math at the University of Maryland. He recalled with pride that Sergey was interested in not only computers but also mathematics. "He was very good at math," said his father. "In middle school, they brought in a special teacher for him and some other students who were advanced."

The family lived modestly in Prince George's County, a suburb just outside Washington DC, and Sergey attended the public Eleanor Roosevelt High School, a tough place where brawn mattered more than brains. One of his classmates recalled that Sergey was "quite cocky about his intellect," often attempting to prove to teachers that they were wrong.

In truth, Sergey didn't think much of many of his teachers, or of his fellow students, and he always felt he learned more at home. While still in high school, he enrolled at the University of Maryland. Studying at an accelerated pace, the 19-year-old whiz received his undergraduate degree in 1993 with honors in math and computer science.

"I got a lot of attention, a lot of one-on-one. I was better prepared than peers from MIT and Harvard," Sergey said of his undergraduate education at Maryland, where he took many graduate-level courses.

"My colleagues said he was a good student," Michael Brin said of his son. "Overall, there's a lot of freedom for good students. They can take quite a variety of courses and also have a lot of qualified faculty."

During the summers, Sergey worked on new analytical tools in a range of disciplines, including developing a 3-D graphics routine suitable for a flight simulator. His summer jobs at Wolfram

Research, General Electric Information Services, and the University of Maryland Institute for Advanced Computer Studies took him further into the computer science, data mining, and mathematics fields.

Sergey also inherited a sense of humor from his parents. On her own Web page, his mother posted a photograph of herself beside a cutout of the Russian Communist dictator Lenin; the caption read "A picture of me with my best friend." When Sergey celebrated a birthday in his twenties, his father wrote him the following poem and posted it on the Internet.

You are growing stronger
In body, spirit and mind.
I am getting older
Leaving decades behind.

You are tough, you mine data,
You surf first and think later,
And your crawler fast as light
Wanders madly in the night.

You work hard to squeeze a thesis
From the world wide web of feces.
You live abroad on the sunny coast
To you, my son I propose a toast.

In the stately oval office
Clinton grimly counts his losses,
Plumpy interns taking turns,
Many wonderful returns.

Michael Brin kept his family, and his students, entertained and on guard. He would often hand graded tests back to his math students with a simple, "My sincere condolences," or amuse himself by making fun of wrong answers, saying, "That's perfectly incorrect." A former student recalled his style as engaging but intimidating: "Dr. Brin is a first-rate orator who begins every class with a smoke. Well, usually he'd show up, pose a rather vexing problem, and then escape for a cigarette in the midst of the ensuing confusion. He expected an answer from the class by the time he returned. . . . Almost *half* the statistics class he taught dropped out after the first session because they couldn't handle his assault and battery on their sense of self. Attending his classes was like experiencing the drill sergeant in *Full Metal Jacket*."

Every member of the Brin family had a homepage on the Internet that was linked to one another, a sign of their connectedness in cyberspace. On his homepage, Michael Brin wrote, "Sergey is a grad student at Stanford (computer science). He does data mining and (with his friend Larry) developed a search engine GOOGLE which he claims is the best." Sergey's much younger brother, Sam, revealed his passion for hoops on his site: "My whole life is based around basketball. I practice every day for at least thirty minutes plus one hour of team practices on Monday and Thursday. My favorite team is the Washington Wizards because they play closest to where I live."

After graduating from Maryland, Sergey went to Stanford as the recipient of a National Science Foundation graduate fellowship, with a focus on computer science. Michael Brin sensed that his son was likely to embrace the academic life, like his father and grandfather before him.

"I expected him to get his Ph.D. and to become somebody, maybe a professor," Michael Brin said. "I asked him if he was taking any advanced courses one semester. He said, 'Yes, advanced swimming.'"

Before he met Larry in 1995, Sergey pursued a variety of subjects that intrigued him at Stanford. He didn't need to load up on graduate-level courses, since he already had mastered many of the subjects at Maryland. So he learned to sail and developed a love for the trapeze. His open-mindedness about exploring new academic subjects also led to a variety of accidental but significant discoveries. He collaborated with fellow Ph.D. students and professors on a project involving automated detection of copyright violations and on a study dealing with molecular biology, and he was excited about an idea for generating personalized movie ratings. "You rate the movies you have seen," he explained. "Then the system finds other users with similar tastes to extrapolate how much you will like some other movies." Brin was onto something; a similar idea for rating books soon became popular on Amazon.com.

Sergey found Stanford an intellectual feast of opportunity for the curious. "I tried so many different things in grad school," he said. "The more you stumble around, the more likely you are to stumble across something valuable."

After Larry Page's arrival in the fall of 1995, he and Sergey began hanging out and working together. Brin left his movie idea behind to pursue other projects that complemented some of the work Page wanted to do. And they began satisfying their shared curiosity about the emerging wonders of the Internet.

As they tinkered in the ivory tower, the world around them was changing radically. From nearby Silicon Valley to Wall Street, the buzz was all about the initial public stock offering of a 16-month-old technology start-up called Netscape. On August 9, 1995, Netscape went public at a stock price of $28 per share and soared to a peak of $75 on the first feverish day of trading. Netscape, a high-tech marvel, suddenly was worth more than $3 billion. The Netscape IPO symbolized the ushering in of the Internet era in Silicon Valley and created a gold-rush mentality. Wall Street was ready. Stockbrokers with no idea what the company did called investors and told them the smart money was betting that this was only the start. It didn't seem to matter that Netscape wasn't generating profits, since sales, though small, were growing 100 percent every quarter. With a nifty product called a browser that enabled computer users to navigate the Internet, the profits would come later. Some analysts even predicted that Netscape would eclipse mighty Microsoft. In fact, before the end of 1995, it hit a price of $171 per share, propelling Wall Street financiers to prowl for other Internet companies with stories to tell and stocks to sell.

After Netscape, the smell of money permeated Stanford's

computer science department. The university, it turned out, did not see a conflict between academics and financial rewards. Its primary mission was to train the next generation of professors and academic researchers, but it had also established itself as an incubator for some of the world's most successful technology companies, from Hewlett-Packard to Sun Microsystems ("Sun" stands for Stanford University Network).

Unlike MIT and some of the other leading research institutions, Stanford made it extremely easy for students in its Ph.D. program to work on potential commercial endeavors using university resources. Its Office of Technology Licensing also took a broad view of its role. Rather than laying claim to all the groundbreaking on-campus work of its students and professors, the office assisted with, and paid for, the patent process, and then entered into long-term licensing agreements that enabled Stanford's rocket scientists to launch start-ups and get rich. In return, the Stanford licensing office often received stock in the new technology companies.

"I never want that to become a barrier to technology transfer," said Stanford's president, John Hennessy. "We have an environment at Stanford that promotes entrepreneurship and risk-taking research. You have this environment that gets people thinking about ways to solve problems that are at the cutting edge. You have an environment that is supportive of taking that out into industry. People really understand here that sometimes the biggest way to deliver an effect to the world is not by writing a paper but by taking technology you believe in and making something of it. We are in an environment where a mile from campus they can talk to people who fund these companies and have lots of experience doing it."

On nearby Sand Hill Road, a number of the nation's most aggressive investment firms bet money on start-ups in return for stock. Known as venture capital firms, they made high-risk investments in early-stage companies in the hope of reaping fat returns. Since nobody had a crystal ball, some bad investments were an inevitable part of the venture capital game. But the best firms had plenty of money to pump into fresh ideas and innovation in the hope of hitting some blockbuster paydays when private companies went public or were sold. The presence of venture capitalists in the neighborhood made it easier for students and professors at Stanford to get funding and advice than for their peers at any other university. By allowing faculty to own a stake in companies and cash in from time to time, Stanford also retained many of its most accomplished professors. Some of them became multimillionaires, and had fun in the process. Given the sunny weather and palm trees, the brilliant students, and the freedom to explore lucrative, cutting-edge ideas, there was no compelling reason for most professors to leave. It was far more interesting than sitting on the beach or working in the private sector.

Larry and Sergey—the sons of professors who conducted research and taught in more traditional academic settings—were focused on pursuing their Ph.D.'s, not on getting rich. In their families, nothing trumped the value of a great education. Beyond taking pride in the paths their parents had pursued, both looked forward to becoming Stanford Ph.D.'s someday, and wearing that badge of honor. Neither of them had the slightest idea just how soon their heartfelt commitment to academia would be tested.

CHAPTER 3
Learning to Count

In January 1996, Larry and Sergey, along with the rest of Stanford's computer science students and faculty, moved into a new home: a handsome four-story building of beige stone with chiseled letters that said WILLIAM GATES COMPUTER SCIENCE. The Microsoft chairman donated $6 million toward its construction, enough to snare the naming rights. While Gates did not attend Stanford, Microsoft hired many of its graduates, and Gates hoped that having his name over the entrance would boost the company's chances of luring top talent in the future. Gates said he made the gift to "invest in the future of the industry." At the dedication ceremony, James Gibbons, dean of the engineering school, predicted that within 18 months "something will happen here, and there will be some place, some office, some corner, where people will point and say, 'Yeah, that's where they worked on the blank in 1996 and 1997. And you know, it was a big deal.'"

Assigned to a third-floor office, Larry Page moved into Gates 360, together with four other graduate students. One of them, Sean Anderson, was creative, eccentric, and intense. He would eventually give up his apartment and begin living and sleeping in the office. Another, Ben Zhu, rarely spoke. Then there was Lucas Periera, a bundle of energy, "very, very, very, very bouncy," according to Tamara Munzner, the only female in the bunch and a self-described geek. The financial boom in Silicon Valley set off by the Internet and the Netscape IPO made loyalty to academics challenging. "It was a hard time to stay in grad school," Munzner said. "Every time you went to a party, you had multiple job offers and they were all real. I had to redecide every term not to leave."

Despite the close quarters, or perhaps because of them, camaraderie developed. Sergey Brin had been assigned to another office, but he quickly began spending much of his time in Gates 360, alongside Page. Stanford officials had asked Brin to devise the numbering system for the offices in the new building. He did the job, and in return, he insisted that the university provide ergonomically better chairs than it had picked initially. "Sergey is very clever," Anderson said.

By then, Gates 360 resembled a small jungle, with hanging plants and vines growing on the ceiling, thanks to Anderson, who put plants on the desks, too. He also brought in a five-gallon bucket of water and a pump controlled by his computer. "I set up an automatic watering system," he said. "We had lots of little toys in the office." He also plugged a piano into his computer that he let everyone use, and Munzner brought in a pad in case anyone wanted to take a nap on the floor.

Larry and Sergey were always together. On campus, they became known as LarryandSergey, a phrase that rolled off the tongue. "They were fun guys to share an office with," Munzner said. "We all kept crazy hours. I remember once at three in the morning on a Saturday night, the office was full. I remember thinking, 'We are such geeks.' We were all very engaged in what we were doing and all pretty happy."

The banter between Larry and Sergey knew no limits. Munzner said they were "goofy-smart" but not conceited. They relished challenging and debating each other and anyone else they could suck into a good argument. They talked endlessly about computers, philosophy, and whatever else popped into their minds. Once they argued loudly about whether it was possible to construct a building-size display out of lima beans, prompting Munzner to whirl around in her chair and exclaim, "You have got to be kidding me!" In a corner of Gates 360, under Larry's desk, they built a computer rack out of Legos. The others in the office found it virtually impossible to get any work done without tuning them out. "I learned to program with headphones," Munzner said.

One of Page's favorite topics of discussion was devising new and better transportation systems. Growing up near Detroit had prompted him to ponder creative ways of moving people and goods from place to place, while cutting down on accidents, expenses, pollution, and traffic. "He liked to talk about automated automobile systems where you have cars that will roam around, and if you need one, you just hop in and tell it where it needs to go. It is like a taxi, but it is cheaper and packs itself with other such vehicles on the freeway much tighter," Anderson recalled. "He is passionate about the problem of moving people or things around. He liked solving the problems of society in various ways."

Rajeev Motwani, a 30-year-old professor who had been Sergey's advisor since his arrival at Stanford in 1993, watched the intellectual kinship between Brin and Page take hold, and he grew increasingly fond of them. "They were both brilliant, some of the smartest people I have ever met," Motwani said. "But they were brilliant in different ways." Sergey was practical, a problem solver, an engineer. If something worked, it worked. He was also mathematical, lightning fast, and outgoing. "He was the brash young man, but he was so smart, it just oozed out of him," Motwani said. Whenever he needed or wanted something, Brin would just barge into Motwani's office without knocking. Page, on the other hand, was a deep thinker. He wanted to know why things worked. Possessed of boundless ambition, Page had a more reserved demeanor. He would knock before entering Motwani's office. "If there was a group meeting of 20 people, Sergey was holding court. You wouldn't notice [Larry] if he was in a crowd, but then afterward he would say, 'Hey, what do you think of this idea?'"

Sergey's manner differed from that of typical Stanford Ph.D. students. "He carries himself with a lot of intensity. He is very direct. The intensity is not common," said Dennis Allison, another Stanford professor. "Sergey is very present. He is able to convey the fact that he is aware of you, talking directly at you and wanting to interact. It is exciting to talk with him."

Brin had been working closely with Motwani on finding ways to extract information from large mountains of data. They had started a new research group called MIDAS; the acronym stood for Mining Data at Stanford. (In Greek mythology, Midas was the king whose magic touch turned everything to gold.) Brin lined up the weekly speakers and chose the topics for MIDAS discussions. He and Motwani also teamed up and wrote a number of papers together on the subject.

While data mining can be used to determine what combinations of items customers purchase in stores so that retailers can arrange products better, Brin and Motwani experimented with applying the same techniques to the emerging, disorganized Internet. In the mid-1990s, the Web was a virtual Wild West—unregulated, uninhibited, and unruly. Millions of people logged on and began communicating via email, but serious researchers grew frustrated amid the clutter of Web sites. Early efforts to help computer users find information on the Internet, including WebCrawler, Lycos, Magellan, Infoseek, Excite, and HotBot, fell short. "Search was not pretty in those days," Motwani said. "You'd get a slew of results that were completely meaningless." Motwani had tested a search engine called Inktomi in 1995 after it was developed at Berkeley, where he had received his Ph.D. He typed in "Inktomi" to see what would happen. Sure enough, said Motwani, "It wasn't there. It couldn't find itself."

Meanwhile, Stanford doctoral candidates Jerry Yang and David Filo took a different approach to search. Rather than relying on technology alone, they employed a team of editors who selected Web sites for an alphabetized directory. They called their company Yahoo! Although their approach simplified finding valuable information, it was not comprehensive and it could not keep pace with the fast growth of the Web. Brin and Motwani also tried other directories and search engines, but nothing got the job done. Instead, a simple search would yield hundreds or thousands of results in no discernible order. It took them hours to sift through the pages manually to find whatever they were seeking. Brin and

Motwani became convinced that there had to be a better way to search the Internet.

At the same time, Page—who had been spending time on something called the Digital Libraries Project—began hunting around the Web using a new search engine called AltaVista. While it returned somewhat better and faster results than the other search engines, Page noticed something else entirely. In addition to a list of Web sites, AltaVista's search results included seemingly obscure information about something called "links." Links contributed to the Web's dynamism; computer users seeing a highlighted word or phrase could click on that link if they wanted to learn more, and they would instantly be taken to another Web page. Instead of focusing on AltaVista's main search results, Page began pondering what could be gleaned from analyzing the links.

Hector Garcia-Molina, one of Page's advisors, agreed that analyzing data about the links was potentially valuable. AltaVista didn't appear to be doing anything with the links other than reporting on them in raw form. Page wanted to dig into links and see how they might be used further. But to test any of his theories, Page would need a big database. Not lacking ambition, he quickly did some calculations and then told his startled Stanford advisor that he was going to download the entire World Wide Web onto his desktop.

On its face, Page's idea seemed more absurd than audacious. He even declared that downloading the Web could be done fairly easily and quickly. Garcia-Molina and others scoffed. Page, however, was deadly serious, and on a mission to capture nothing less for his research. He was in good company. Tim Berners-Lee, the British computer scientist who invented the Web in 1989, had the visionary notion that a click on highlighted words would propel information-hungry computer users from one document to another to another. To a visionary computer maven, the Web was all about links.

As 1996 wore on, Page and Brin teamed up on the work of downloading and analyzing Web links. It took longer to get the data than Page had envisioned—he estimated that it cost the computer science department $20,000 every time they dispatched a "spider" program to canvass the entire Internet—but he desperately wanted to see it through. His drive to discover the importance of these automated cross-references attracted the attention not only of Brin but also of Brin's advisor, Motwani, since it held out the promise of improving Web research. Brin was drawn to the project by the chance to work with Page and by his own interest in extracting information from giant amounts of random data. What could be bigger or better than the entire World Wide Web for Brin to attack with his math and programming skills?

Page had a theory. Counting the number of links pointing to a Web site was a way of ranking that site's popularity. While popularity and quality don't go hand in hand, he and Brin both had grown up in homes that valued scholarly research published in academic journals with citations. The links, in a sense, reminded Page of the citations. Scientists would cite the published papers their work drew upon, and these citations were a helpful way of tracking credit and influence in the academic and research communities. "Citations are important," Page said. "It turns out, people who win the Nobel Prize have citations from 10,000 different papers." A large number of citations in scientific literature, he said, "means your work was important, because other people thought it was worth mentioning."

The same could be said for Web sites, Page concluded. Taking things a step further, he hit upon a conceptual breakthrough: All links were not created equal. Some mattered more than others. He would give greater weight to incoming links from important sites. How would he decide what sites were important? The sites with the most links pointing to them, quite simply, were more important than sites with fewer links. In other words, if the popular Yahoo homepage linked to an Internet site, that site instantly became more important. Playing off his own last name and the Web documents he was scouring, Page began calling his link-rating system "PageRank."

Another of Page's advisors, Stanford professor Terry Winograd, said that the intellectual path toward discovering how to rank Web pages ultimately revolved around a "one-shot idea": the notion of what could be learned from tracking links. "Larry talked about the idea initially as random surfing, the idea of a random walk on the Web. The motivation for the algorithm [a set of mathematical equations] was really thinking about the surfer. Start on a page, click on a link, and see where you would land most of the time. That got refined into PageRank."

Brin and Page were persuaded that they had found the path toward a Ph.D. thesis by applying PageRank to the Internet. By early 1997, Page had developed a primitive search engine that he named "BackRub" because it dealt with the incoming—or "back"— links to Web pages. Ever thrifty, Page put his left hand on a scanner, converted the image to black and white, and the new BackRub site had its logo. Page, Brin, and Motwani all contributed ideas to the evolving project. Motwani said that it would soon become clear that what they had created together was more than just a way to further their academic research. Without intending to, the trio had devised a ranking system for the Internet, and in the process had inadvertently solved one of the core problems of searching for information on the Web.

"It wasn't that they sat down and said, 'Let's build the next great search engine.' They were trying to solve interesting problems and stumbled upon some neat ideas," Motwani said. "Larry added ideas, Sergey added ideas, I added ideas, we all did; and it became clear we could build a full-scale search engine." Winograd agreed about the evolution of the concept. "They didn't set out to build a company, but they did set out to do better search."

Brin, Page, and Motwani put together a prototype of a comprehensive search engine for internal use at Stanford. Based on conventional search engine technology, with the addition of PageRank, it was a way to search the Internet for information that returned prioritized results based on relevance. While other search engines relied on matching words in queries with words on Web pages, PageRank provided an extra dimension: it put search results in a logical order for computer users. For the first time, there was a way to do an Internet search and find useful answers swiftly.

In the fall of 1997, Brin and Page decided that the BackRub search engine needed a new name. Page was having trouble coming up with a catchy name that hadn't already been taken, so he asked his office-mate Sean Anderson for help. "I would go to the whiteboard and start brainstorming and he would say, 'No, no, no,'" Anderson recalled. This went on for days. "He started getting desperate, and we had another brainstorming session. I was sitting at the whiteboard and one of the last things I came up with was, 'How about Googleplex?' I said, 'You are trying to come up with a company that searches and indexes and allows people to organize vast amounts of data. Googleplex is a huge number.' He liked that. He said, 'How about we try Google?' He liked it shorter. I typed in G-o-o-g-l-e and misspelled it on my workstation, and that was available. Larry found that acceptable, and he registered it later that evening and wrote it on the whiteboard: Google.com. It had a wild Internet ring to it, like Yahoo or Amazon. And I came in the next morning and Tamara had written a note saying, 'You misspelled it. It is supposed to be G-o-o-g-o-l.' Of course that was already taken."

In 1997, the search engine was made available internally to students, faculty, and administrators at google.stanford.edu. Around the campus, its popularity grew by word of mouth. The university's Office of Technology Licensing sought a patent. And Stanford professors and students began using it to hunt for information online. "It instantly became my only search engine," said Stanford professor Dennis Allison. "Google became my default right away," added Winograd. "It spread through other parts of Stanford."

Lacking the funds to hire a designer and the artistic talent to create something elegant, Brin kept the Google homepage simple. From the start, Google's clean, pristine look attracted computer users hunting for information. In a cluttered world, its primary colors and white background conveyed purity, with universal appeal. It stood in marked contrast to the growing number of busy-looking Internet pages with flashy ads and crowded graphics and type. Because it didn't feel as though Google was trying to sell anything, people took ownership of the search engine and more readily adopted it as their own. "As a piece of design, it is really brilliant," Allison said. "If you went to a design firm and asked for a homepage for a search engine, you would never get that. It doesn't have any animation or metallic colors, and there is no sound or lights. It flies completely in the face of the common belief that people love to find their way through the noise."

As the database and user base grew, Brin and Page needed more computers. Short of cash, they saved money by buying parts, building their own machines, and scrounging around the loading dock looking for unclaimed computers. "We would just borrow a few machines, figuring if they didn't pick it up right away, they didn't need it so badly," according to Brin. Their advisors, who knew of their scavenging, also funded them with $10,000 from the Stanford Digital Libraries Project. After cramming as many computers as they could into Gates 360, they turned Page's dorm room into a data center. "We assembled quite a mishmash of things," Brin said, noting that they learned a valuable lesson: how very much they could accomplish by assembling and stringing together inexpensive PCs. "Larry would scour the world to save a penny," said Charlie Orgish, Stanford's head of computer systems.

Seated in Palo Alto's Mandarin Gourmet restaurant in March 1998, Page and Brin prepared to pitch Paul Flaherty, a Stanford Ph.D. and an architect of AltaVista, on the merits of their superior search engine technology. AltaVista, they hoped, would pay as much as $1 million to get access to the soon-to-be-patented PageRank system. After all, it would improve their search results, and given AltaVista's 54 percent market share in search, it seemed logical that the company would want to implement the best technology. Brin and Page would then be able to resume their studies at Stanford.

After listening to Flaherty offer an explanation of how AltaVista worked, the Google Guys knew they had something better. Still, one of the things Flaherty said lodged in their minds—that the entire AltaVista database, if printed out, would amount to a stack of paper 60 miles high, and that the search engine could pull any word out of that stack in less than half a second. The image stuck. Dennis Allison, the host of the dinner, wondered if there would be fireworks or friendships by the time the fortune cookies arrived, given how strongly Brin and Page came across.

AltaVista, they said, was just the beginning; Google was the future. Flaherty agreed that the guys had a cool concept. "I felt they really had something with their link-based approach to page ranking, which was AltaVista's technical weakness at the time," Flaherty said. But he also sounded a note of caution. Problems and headaches happen on the Internet after you become popular, he told them. People try to break into your network, attack your site, and manipulate your search system. But Page and Brin weren't afraid; instead, they brimmed with confidence and wanted their technology in the hands of more people. "They were excited by what they were able to do," Allison said. "They were very anxious to share it. They were basically saying in a polite way, 'AltaVista is dead meat. It doesn't do what is needed.' These guys had a PageRank."

But a few weeks after their Chinese dinner, Brin and Page heard back from Flaherty that AltaVista was taking a pass on Google. Its parent company, Digital Equipment Corp., didn't like relying on outsiders. "The people who were running engineering weren't very open to outside technology," Flaherty said. "They had a big 'not invented here' attitude." Also complicating matters was Digital's pending merger with Compaq Computer. The search engine simply wasn't a focus or priority, especially since AltaVista was moving toward becoming a one-stop destination for people to begin their online experiences. Search was only one of numerous offerings that AltaVista would provide Internet users in addition to news, shopping, email, and more.

With the help of Stanford professors and the Office of Technology Licensing, Brin and Page tried unsuccessfully to sell their PageRank system to Excite and other search engines. It didn't seem to matter that they had something better. Everyone around them seemed to be focused on selling as many ads as they could to cash in as fast as they could. Winograd accompanied the pair on a visit to a venture capital firm on Sand Hill Road, but nobody was excited about funding "search." While Larry and Sergey saw the search engine as special and the most important part of the Internet experience for computer users hunting for information, others saw it as a sideline, merely one of a number of items to be included in a smorgasbord of services. But the pair didn't give up. "They have a somewhat skeptical view of authority," Winograd said. "If they see the world going one way and they believe it should be going the other way, they are more likely to say, 'The rest of the world is wrong,' rather than, 'Maybe we should reconsider.' They were confident in their approach and would tell you they thought everyone else was wrong."

Yahoo, seemingly a logical buyer because it relied on directories edited by people and didn't have a fast way to scour the entire Internet, also turned down the chance to buy or license the Google technology. In part, Yahoo rejected it because the firm wanted computer users to spend more time on Yahoo. The Google search engine was designed to give people fast answers to their questions by swiftly sending them to the most relevant Web site. The Yahoo directories were designed both to answer questions and to keep people on the Yahoo site, where they could shop, view ads, check their email, play games, and spend more money and time, rather than less. Yahoo cofounder David Filo advised Brin and Page that if they wanted to realize the potential of their unique search system and believed in it, the best thing for them to do was to take a leave of absence from the Ph.D. program at Stanford and start their own business. That way, he said, they could design a business of their own that fit with their search engine. If it was as good as they claimed, it would catch on, since Internet use was growing rapidly and computer users gravitated to the best services and sites.

The rejections frustrated and upset Larry and Sergey, but also increased their determination. They didn't know what to do. "They were ambivalent," said Winograd. "'Are we leaving? Are we doing this on the side?'" After many months of being turned down, they decided that, at least for the time being, they would focus on improving Google as much as possible for Stanford users, and put off any major decisions. Tinkering one day with a graphics program called GIMP, Sergey created a color rendering of the Google letters with an exclamation point at the end, mimicking Yahoo! He seemed quite proud of the new logo, which was composed of kindergarten-style block letters in primary colors. But it wasn't the look that meant the most to him. He was pleased that he had been able to teach himself how to use GIMP, free software that was tricky to employ.

In the spring of 1998, Brin and Page sent out an email newsletter to a list they called Google Friends and urged people to spread the word. "Google has now been up for over a month with the current database and we would like to hear back from you," they wrote. "How do you like the search results? What do you think of the new logo and formatting? Do the new features work for you? Comments, criticisms, bugs, ideas, . . . welcome. Cheers, —Larry and Sergey."

By July they had added a summary, or snippet, for each search result. This highlighted in bold letters the parts of a Web page that were specifically responsive to a query. This step meant Google users could see which results best answered their question without having to visit several sites. "Expect to see a lot of changes in Google in the next few months. We plan to have a much bigger index than our current 24 million pages soon. Thanks to all the people who have sent us logos and suggestions. Keep them coming. Have fun and keep googling," they wrote.

Despite the cheery emails, Winograd knew that Larry and Sergey had hit a brick wall. To really grow Google, they needed to move off campus and take some risk. But without funding, they didn't have a way to buy the computer parts they needed to give it a try. Winograd empathized with Page's predicament. "I said, 'I don't see how you're ever going to get the money.' He'd say, 'Well, you're going to see. We'll figure that out.'"

CHAPTER 4

The Secret Sauce

On a sunny California morning in late August 1998, Larry and Sergey sat on the front porch of a Palo Alto house eagerly awaiting the arrival of a Silicon Valley angel. The pair had been working day and night on their new way to search the Internet, but despite being economical and saving money wherever possible, they had run out of cash. One of their graduate school professors, David Cheriton, suggested that it would be a good idea for them to meet his friend Andy Bechtolsheim, a computer whiz and legendary investor in a string of successful start-ups.

Bechtolsheim pulled up in front of Cheriton's house in his silver Porsche, hopped out of the car, and made his way to the porch, where the trio was waiting for him. Like a handful of other wealthy entrepreneurs in Silicon Valley, Bechtolsheim worked hard, even though he didn't need the money, because he was passionate about the power of technology and loved finding new ways to solve problems.

He was also quite modest. Many of those who worked on technology projects with Bechtolsheim at Cisco Systems, where he served as a vice president, had no idea he was a cofounder of Sun Microsystems and had founded and sold another firm to Cisco two years earlier for hundreds of millions of dollars. Cheriton had piqued Bechtolsheim's interest in Brin and Page by telling him that the students had a "great idea" to show him: they had invented a better way to find relevant information fast on the Internet. He seemed interested. "I used the Internet at the time primarily for searching," Bechtolsheim said, "and part of my frustration was that AltaVista, which was the standard, was not very good."

He wanted to learn more. How far could the guys run with the idea? Page confidently told Bechtolsheim that they could download, index, and rapidly search the entire Internet using a network of low-cost personal computers. The only problem was that they didn't have the money to buy the machines.

Bechtolsheim liked the idea but wondered about its commercial viability, since AltaVista and other search engines on the market were losing money. Some people even saw search engines as commodities, a kind of card catalog for the World Wide Web that had no sustainable edge or enduring competitive advantage. Bechtolsheim wasn't sure if that was the correct analysis. After all, many years before, a single innovation, the Dewey decimal system, had changed the way millions of volumes were organized in libraries around the world.

Sitting on the porch that morning, Brin and Page felt at ease with Bechtolsheim, and he liked their moxie. During the course of his hectic days, Bechtolsheim regularly learned about plenty of gee-whiz technologies that didn't translate into terrific businesses. Throughout the high-tech boom taking place around him, as new companies with ".com" in their names seemed to go public every 15 seconds, Bechtolsheim retained a healthy skepticism for fancy PowerPoint presentations that wowed others. He had a way of keeping his head about him by proceeding with simple methods of evaluating new ventures. He didn't like to bet on promises. Instead, he looked for several things: ideas that solved real problems he could understand; businesses with the potential to produce real profits; and bright, passionate, and capable founders.

He also trusted his instincts and a handful of Silicon Valley colleagues. Cheriton, a professor with business experience and savvy, was among this coterie of technology experts. Cheriton, in turn, knew that Bechtolsheim's involvement with this or any other venture could dramatically increase its prospects for success, not only because of his financial backing but also because of his connections with the leading money men and technologists of Silicon Valley, as well as the rigorous way he scrutinized new ideas and young talent. As Bechtolsheim engaged in dialogue with the students that morning, he could immediately tell that they were intelligent and driven, even if they were short on money and experience.

After Brin and Page displayed the demo and chatted with him, Bechtolsheim appreciated and understood the breakthroughs that enabled Google to produce superior search results. He also admired something else about Brin and Page. Instead of wasting lots of money on advertising or high-end equipment, they wanted to buy motherboards and other components to inexpensively build computers themselves. They also wanted to develop a fully searchable database before going out to talk to venture capital firms about possible funding. And they wanted to let their search engine speak for itself.

"Other Web sites took a good chunk of venture funding and spent it on advertising," Bechtolsheim said. "They believed in word of mouth. This was the opposite approach. Build something of value and deliver a service compelling enough that people would just use it."

Satisfied that he had seen and understood a demonstration of a better technology that had the potential to address a real problem, Bechtolsheim wasted no time asking about the bottom line.

"The key question with any Internet start-up is, 'How are you going to make money?'" he said. "I never get sucked into ideas with no economic merit."

He reviewed various possibilities: build an audience of computer users by giving away the Google search engine for free, and then profit from ads or by selling something. Larry and Sergey had an instinctive aversion to advertising, coupled with a deep-seated fear that it would corrupt search results. They had made that clear in a paper they had written about the search engine. They also talked with Bechtolsheim about licensing search technology to companies that would pay to use it. And there was always the possibility that a large company might buy the technology and include it in its mix of products.

To Bechtolsheim, the search engine conjured up the image of an electronic, searchable directory similar to the yellow pages, where clearly marked display ads appear on the same page as phone listings for plumbers, tennis rackets, or any other product or service.

"This is the single best idea I have heard in years," Bechtolsheim said. "I want to be part of this."

Neither Brin nor Page knew exactly how to respond. Bechtolsheim, who had done this many times before, made it easy for them. He proposed immediately writing a check so they could build their computers and he could be on his way to his next appointment. No negotiations. No discussions of stock or valuation. Bechtolsheim didn't even know that Brin and Page had not formally created a company. But details like those didn't bother him. He never forgot the way an early investor in Sun had once handed him a check on the spur of the moment. It gave that individual immediate involvement in what turned out to be an extremely valuable enterprise. He wanted to do the same thing with Google. Instead of discussing all the details, Brin recalled, Bechtolsheim wrote a check made out to "Google Inc." for $100,000, a figure he picked because it was a nice, round number.

Page put the check in his desk drawer for safekeeping, where it remained for two weeks until the pair incorporated Google and opened the first bank account in the newly established company's name, so they had somewhere to deposit it.

Cheriton's matchmaking that morning on his porch had been a success. The pair of twentysomethings were so excited that they went off to celebrate by eating at Burger King. The endorsement from Bechtolsheim had also given them the confidence and credibility they needed to seek money from family and friends. Altogether, in short order, they raised about $1 million, enough to buy the computer equipment they needed and take the important next steps on their project.

Andy Bechtolsheim, the man with the Midas touch, sped away

in his Porsche that morning without understanding the true significance of what he had just done. "In the back of my mind, I thought maybe they could get millions of people searching and add it all up and make money," he said.

"I didn't know how big this could be at the time. Nobody knew."

Behind closed doors in room 380 of the Gates Building on the Stanford campus, Sergey Brin and Larry Page were getting ready to give Google a trial run before the gathered intellectual elite. It was September 1998, and they had been invited by Stanford professor Dennis Allison to discuss the Google search engine they had developed with the help of others, and its superiority over better-known technologies for finding information on the Internet. The assembled graduate students and computer science professors waited with anticipation. At the front of the classroom, Brin and Page had decided that they didn't want to speak of money at all. This was the time to talk about concepts, the opportunity to formally explain for the first time—in a large setting at Stanford—how they had blended various ingredients in the secret sauce to make the Google search engine the best and fastest way to find information on the Web.

Their talk was part of an ongoing Wednesday speaker series that Allison hosted about important advancements and breakthroughs in technology. His immense respect for Brin and Page was based on their lively personalities, their intellectual horsepower, the unusual maturity they displayed despite their youth, and an ambition that translated into a willingness to tackle subjects others had found too daunting. To Brin and Page, the vast challenges were invigorating. While Allison had seen many brilliant computer scientists, mathematicians, and technologists come and go over the years, there definitely was something special about these two. He had watched as the founders of Sun Microsystems, Yahoo, Logitech, and others marched through Stanford's hallowed halls. The Google Guys were different.

"They are extraordinary people," Allison said. "They embody the best of computer hacking."

By using the word *hacking*, Allison was not referring to the criminal practice of breaking into computer networks or disrupting Web sites. It was, in the world of software engineering and computer science, a term of art describing their ability to write software that truly broke new ground and opened up new possibilities for innovation. And it was a critical difference. Plenty of people around Stanford had great ideas, and many were highly motivated, but few were able to execute their ideas and overcome the hurdles stacked against them. What he admired was the way they held fast to their bold dreams of changing the world.

"They are really driven by a vision of how things ought to be, and not to make money," Allison said. "The idea of digitizing the entirety of the universe and making it work is something nobody was willing to tackle but lots of people knew needed to be done. They managed to get that together and bulldozed through the limitations. And with some luck, it is actually going to work."

For Brin and Page, Ph.D. students who had one foot in academia and the other on the way out the door to launch a company, this was their major opportunity to discuss and present their findings to a large group and get feedback in the academic environment where the concepts had been hatched. They had been warned by many people that once they left Stanford to start their company, secrecy would be paramount to prevent competitors from learning about their performance, their strategy, and their trade secrets. Even within room 380, there was only so far they were willing to go. With so many smart minds listening to their every word, they had no desire to breed a competitor.

"I'm Sergey Brin and that's Larry over there. I'm gonna jump right into it here," Brin began in his typically informal style. He opened with an example of a person going on a trip who typed the words *rental car* into another search engine. The problem with the results, he explained, was that they were ranked solely by counting the number of times the words *rental* and *car* appeared on various Web pages. In contrast, typing *rental car* into Google would give enhanced search results based on relevance. With a savvy audience watching, Brin typed the term into Google and showed that the top results were the official Web sites for Avis, National, Dollar, Alamo, and others, swiftly plucked from the millions of pages on the World Wide Web.

"What we've tried to do with our search engine is not give outrageous answers based on the count of words in the text," Brin explained. "The research started something close to three years ago. That research basically led to innovations about how important a Web page is and the PageRank algorithm, which is a major component of the Google search engine that I'm gonna tell you about today."

While it was easy to talk about the way Web sites linked to one another, it was more complex to derive meaningful search results by somehow taking these links into consideration, along with dozens of other factors. But that was what it took to devise a better search engine. To get there, one had to think about the motivations of the Web site operators who established the links in the first place: they sought to direct computer users to valuable information elsewhere on the Internet and hoped that this, in turn, would bring more visits to their own sites.

"Let me tell you what the challenges are of a search engine," Brin said. "You have to index the entire Web. Quite a bit of data. We'll show you how to manage it. We'll tell you what it is we do to generate better search results, what we expect to get into three years down the road, and the social issues also."

For those with the inclination and time to get into more technical explanations, Brin invited them to sign up for a course that he and Page were teaching that semester, on search engines, and promised the students access to "resources which you cannot find anywhere else in the world." For the time being, however, Brin and Page would use the next hour to do their best to explain the basics of how they developed the Google search engine.

"What is it," Brin said, "that allows you to build a search engine in the first place?"

It was a rhetorical question, but just in case someone in the room didn't quite understand that, he continued without skipping a beat.

"There are millions of Web sites out there, hundreds of millions of Web pages—at last count earlier this year, we were at 300 million Web pages," Brin said. "So how do you do this? Well, it turns out that it isn't so bad. On our side, we have Moore's Law."

Nothing was more important or critical to innovation than understanding the basics of Moore's Law. In the mid-1960s, an engineer named Gordon Moore, one of the founders of Intel, predicted that there would be a doubling in computing power every few years. Without this, the trend leading from the giant supercomputers that were the province of corporations and universities and governments would never have led to the invention of the personal computer, which empowered the individual in dramatic fashion. The trend he predicted occurred over and over again as the years passed. What Brin was explaining now was how Moore's Law applied to building a search engine to survey the entire Internet.

"At the same time that people are generating all this text and stuff, disks are getting a lot faster. We will be able to put all human knowledge, and any information people generate, into your pocket, excluding video feeds, in the next couple decades," Brin said confidently. "All this stuff is doable. You keep it all in one central place. You can process over this stuff, which is what we've done."

The problem, Brin said, is that people can't process thousands of search results. "Humans, unfortunately, are not subject to Moore's Law right now. They've been evolving a little bit slower. That's a big problem. Really, we need to address that." The audience laughed.

Having received the $100,000 check from Andy Bechtolsheim, and having made the decision to take a leave of absence from the Stanford Ph.D. program, Brin wanted to let the academic experts know the general direction that he and Larry were heading. "We are in the process of commercializing Google. And you should see it at Google.com sometime in the near future. There are several things that we use to make our search engine better right now. And a few details I will not get into."

Listening to Brin, Allison recognized that, like science fiction, this stuff was easier to talk about than it was to execute flawlessly. But if anyone could do it, Allison sensed, it was the Google Guys—and they were well on their way.

The Google search engine took more factors into account than any other search engine on the market. It didn't just count words or count links and deliver results. It combined information about links and words with other variables, in new and interesting ways that produced better search results. For example, Brin said, it mattered whether words or phrases on Web pages were close together or far apart, what their font size was, whether they were capitalized or in lowercase type.

"We try not to throw away too much information," he said. "We take a user approach. We sacrifice a lot of computing power to produce good results, but that's what it takes."

Doing quality search not only required the right mathematical formulas and equations in the software, but, fundamentally and importantly, it required much greater computing power than any existing search engine on the market had yet employed. It was fine with Larry and Sergey that others overlooked this critical ingredient in the search recipe. But they had understood from the start that the only way to deliver quality search results to computer users was to invest, as nobody had before, in computers. The pair not only planned to develop software, they also intended to focus enormous attention on every aspect of the computer network, because this too was part of delivering the best search results. The software and hardware were inextricably intertwined, and optimizing them was essential. They knew that they would benefit from the falling prices and increasing power of computer memory and other components. They would exploit their innate ability to develop the software, buy the components, and build cheap clone PCs themselves that could handle the task. Their vision encompassed a more holistic approach to delivering high-quality search results than anyone else had conceived of and executed.

As Dennis Allison listened to the talk, he reflected on his experiences with Brin and Page at Stanford. He admired them. They were trustworthy and wanted to do the right thing. Strip away all the technical knowledge and what you found were two young guys with character. That would translate well into the work they did, especially in a field where people needed to trust you to trust your products. They were also, in Allison's view, computer nerds with an unusually wide array of outside interests, a combination necessary for success in the world at large. They were social progressives, which in Silicon Valley meant that they favored free and open software systems, rather than the closed systems that Bill Gates and Microsoft generally preferred. And they were highly opinionated, too.

"They don't like some of the things that happen in corporate America and have not a caution in the world about stating that," Allison said.

As Page took charge of the talk, Allison continued to be impressed. Larry was a very good teacher. He could find the key idea in something and express it in a nontechnical way so that everyone understood. To Allison, that was the mark of a clear-headed thinker who knew his stuff.

"Every time you create a link," Page told the hushed audience, "you've created a citation. But if you just try to count citations on the Web, which is what a lot of search engines do, you run into problems. The Web isn't like scientific literature, because anybody can produce Web pages.

"One way to think of PageRank," he explained, "is it's a usage model. You have a random [Web] surfer. It's kind of like a monkey, just somebody who sits around and clicks links all day and doesn't have any intelligence. You could argue that this kind of approximates people's behavior on the Web." Page paused, the audience chuckled, and he went on.

"PageRank is basically saying, if somebody points to you, you get some fraction of the importance that they have. Let's say somebody really important points to you. That's worth more than someone who has a random Web page. For example, if Yahoo points to you from their homepage, that's a big deal. If you just have one link on the Yahoo homepage, that's pretty good. Either you had to pay someone a lot of money, or your page is pretty good. If you have a link on my homepage, nobody would pretty much care." Page then explained how he derived his recipe for producing ranked search results. "We've assigned numbers to those pages to correspond roughly to their importance. The page's ranking is the sum of all things pointing to it."

Another major challenge for search engines, Page said, is that people try to trick them so that their Web sites get a higher ranking in the list of results. The search engine has to win the cyber war by being smarter than these manipulative Web sites.

"People try to mislead search engines," Page said. "How many times have people seen porn results thrown in with other random things? Raise your hands. Okay, we have a few who admit to it. It is a big problem for search engines. People trying to make money, basically, by getting their results listed for every search query, they don't care what you're searching for, they don't care about anything. They just want traffic on their Web pages. It is a very serious problem."

Having identified the problem, Page explained that he was well on his way to devising the solution. The answer, he said, would involve a dynamic and ever-changing way of measuring the true importance of Web sites. That would make it harder for Web site operators to game the system. With total focus on the end user, Google would get the job done.

At this point, Page couldn't resist taking another shot at what he saw as the sorry state of other search engine technology. "Search engines don't work very well," he said. "If you type AltaVista into another search engine, will you get the AltaVista homepage? Maybe not. That's something we do a really good job at. We do all this stuff ourselves. It is a fairly major undertaking."

The key to Google's approach was breaking down complex tasks into smaller chunks that could be handled simultaneously. Using the right mathematical equations and multiple personal computers, Brin and Page could create a modern assembly line to process the gathering, indexing, and presenting of information, and take advantage of Moore's Law, too, assuring them more computing power for less cost.

"We crawl the Web, which means we go out and download the entire Web. We download roughly 100 pages per second," Page said. "This is fairly complicated to do reliably. We actually store all the Web pages we download because it's very good for research. We have the Web on disks across the hall. It promises to be very useful for research to have this around."

As Larry Page revealed a bit more about what made the Google search engine better than the others, the Stanford students and professors hung on every word.

"Whenever you query with more than one word, we're looking at the distance between words [on a Web page]," he said. For that, a series of equations driving sophisticated software came in handy. Other search engines used a more simplistic approach and failed to keep pace with the growth of the Web. But with robust mathematical formulas at work, downloading as much of the Web as possible and being comprehensive also set Google apart.

"If you want to get more stuff, you should just crawl more of the Web," Page said. "That's the easy way of doing it."

Larry and Sergey were careful not to give away all of the secrets that went into PageRank or Google. There could be spies from other companies in the room, and they didn't want to risk someone ripping off all of their hard work.

At this point, Brin decided to liven the subject. Crawling and indexing the Internet might sound totally technical, he said, but it was also adventurous and even dangerous. From the perspective of some Web site owners, Sergey explained, the crawling spider represented an unwanted intrusion.

"To touch on another aspect of crawling," Brin said, "we have lots of fun. If you are contacting a million Web sites, you are basically contacting a million people who are those webmasters. So imagine going around door to door and knocking on a million homes and giving them your email address. What would be the odds of staying alive during this process, say, in certain parts of Oakland?"

Brin said a small number of "crazy" webmasters, upset that the Google crawler had disrupted their sites, retaliated with massive email attacks or threatened legal action. "They talk to you, try to sue you, and we ended up shutting off Web sites in Montana. At one point, we shut off all of Singapore. . . . Occasionally you get these people and they'll contact the officer of risk management at Stanford—you don't even know such a person existed. Well, we do now. He contacted us. It's no end of trouble."

CHAPTER 5

Divide and Conquer

When Brin and Page took leave from Stanford University in the fall of 1998 to pursue the building of the world's best search engine, they moved their computers, gadgets, and toys into the garage and several rooms of a house with a hot tub in nearby Menlo Park. The house's owner and their first landlord, Susan Wojcicki, knew Brin because he had dated her roommate. Brin and Page could have rented the space for $1,500 a month, but they chose instead to pay $1,700 so that all fees and taxes would be paid, and everything would be done properly from the start. On September 7, they formally established the company Google Inc. They then opened their first bank account and deposited Bechtolsheim's $100,000 check. They also hired a friend, Craig Silverstein, a fellow Stanford Ph.D. student, as their first employee. "We found a place with some extra space and moved in and worked in a garage, like a good Silicon Valley start-up should do," Silverstein said. Wojcicki, though, was in for a bit of a surprise, since she thought the guys in the garage would only be there when she was not. "We thought, 'Well, they'll probably just be there during the day while we're at work. We won't notice,'" Wojcicki said. "But they were actually there 24 hours a day, all the time. But in the end, it worked out well. And we got free Internet access at the same time."

After five months, Brin and Page outgrew Wojcicki's garage, so in early 1999 they moved into offices on University Avenue in downtown Palo Alto. It would be the first of several changes of venue that would serve as markers of the company's emerging culture and growth. They wanted work to be fascinating and fun, and were determined to keep it that way. The second-floor space in the heart of a posh college town was a perfect place to be, barely a mile from the Stanford campus and much livelier than an office park. Neither of the guys had a clear idea of how the company would make money, though it seemed to them that if they had the best search engine, others would want to use it in their organizations. Most of all, they shared an abiding excitement about helping people find more relevant information online, and faster. That remained their primary motivation. "We started this company because we were unhappy with current search technology," Page said. "If we are successful, that will just be a great side effect."

Already quite popular at Stanford and with insiders, their search engine was handling about 100,000 queries per day. This growth had been entirely through word of mouth, emails, and instant messages: free, powerful forms of viral marketing and momentum that they didn't want to lose just because they were now off campus and more isolated. In January of 1999, the pair gave a talk to about 40 students and others at Stanford, and they made it a point to stay in touch with their professors. In February, they sent a newsletter to their users and friends.

"Google the research project has become Google.com. We want to bring higher quality and greatly improved search to the world, and a company seems to be the best vehicle for accomplishing that goal," Brin and Page wrote. "We've been hiring more staff and putting up more servers to scale the system (we've started ordering our computers in 21 packs). We've also begun crawling more often, so our results not only remain fast, they also remain up to date. We are rapidly hiring talented people to bring the latest and greatest technology to the Web."

Brin and Page offered ten reasons to work for Google, including cool technology, stock options, free snacks and drinks, and the proposition that millions of people "will use and appreciate your software." They had set their sights high, despite being rejected early on by some potential business partners and investors, and their enthusiasm continued to mount along with the number of users. Others were starting to take note. Although Google was still in beta—or test—form, it was included in *PC Magazine*'s list of the Top 100 Web Sites and Search Engines for 1998.

A number of converging trends gave Brin and Page a further edge. The better-known commercial search engines, including AltaVista, Excite, and Lycos, were journeying from their core mission of investing in better search technology, either because they were lost inside larger companies or they were chasing the advertising dollars flowing from newly public dot-coms with more cash than smarts. The resulting decline in search quality drove users to hunt for alternatives, and as they did, more and more found their way to Google. The mention in *PC Magazine* put Google on the radar screen of thousands of people for the first time and it also taught the guys a lesson about the power of free media exposure: as their Silicon Valley neighbors blew millions of dollars they didn't really have on Super Bowl ads and extravagant marketing, Google grew in popularity and recognition without spending a dime.

The prevailing wisdom was that all-purpose Web sites would be the preferred gateway to the Internet, but Brin and Page didn't believe it for a moment. By trying to be all things to all people, these portals would end up not meeting any particular or specialized need that would distinguish one site from another, just as the Internet evolved toward personalization. Single-minded and focused, Page and Brin remained convinced that Internet search was the most important long-term problem they could solve, and that doing so would bring new users in droves. To ensure the accuracy, speed, and reliability of every search, the pair had invested most of the money they had raised in additional computer hardware, and most of their time in scouting for talent and improving software. They had a target in mind: becoming dominant in search, at the exact time that others were abandoning it and even derisively calling it a commodity. The two remained steadfast in their belief that search was critical to navigating the expanding World Wide Web. And as the number of queries on Google grew, they also began to recognize that they had inadvertently developed an extremely popular brand name and logo.

Soon after moving into their new offices in Palo Alto, Google expanded to eight employees and struggled to keep up with the growing number of daily searches. Its unique system of computing, which adapted cheap PC parts and custom software into a small supercomputer, gave it the power to handle a rising number of search requests and ever larger downloads of the Web. But there were times when it seemed it would be impossible to keep pace with demand, particularly because Brin and Page were running out of the initial $1 million in funding that they had collected from Andy Bechtolsheim and other early investors, and through borrowing on their own credit cards. As the year wore on and traffic surged to 500,000 searches per day, it was clear to Brin and Page that they had to come up with a substantial chunk of money to keep adding computers to their system. The more computers they added, the more queries they could handle. The more queries they handled, the faster their business could grow. What the two did not want to do, however, was lose control over their company.

In the boomtown climate of Silicon Valley in early 1999, raising money through a public stock offering was one option available to Google even if it had no profits. But Brin and Page had no desire to reveal their trade secrets and methods by going public, and they weren't really interested in the money as much as the chance to continue building their search engine. Rounding up more angel investors would be impractical because of the sizable sum of cash they required. And growing the business themselves appeared unfeasible. They had begun licensing their search technology to companies who wished to have it on their internal or external networks, landing software firm Red Hat as their first official customer. But this was an exception. They found it hard to persuade people to pay for search services when the consensus among businesses was that search did not matter. What they needed was a large cash infusion from outside.

Page and Brin studied the venture capital scene. They were determined to raise the money from a blue-chip venture firm without giving up control. That seemed about as likely as Page's idea to download and search the entire Web had once seemed. But they were supremely confident that they could pull this off as well. Everyone in Silicon Valley knew that it was every entrepreneur's dream to land funding from one of the prestigious venture capital firms on Sand Hill Road. The right money from the right people led to the right contacts that could make or break a technology business. At the same time, giving up control to venture capitalists could destroy the vision of a firm's founders and the long-term potential of a breakthrough technology.

Examining the biggest pools of potential funds to tap, the Google Guys learned how company after company had lost control of their destiny. Venture capitalists had either ramped up the firms quickly for an initial public offering or pushed them to bring in as much cash as possible through advertising. With tech-savvy early investors such as Amazon.com CEO Jeff Bezos to provide them with advice, Brin and Page decided to reach out simultaneously to two of the most established and prestigious venture capital firms in Silicon Valley: Kleiner Perkins Caufield & Byers, and Sequoia Capital. If things went well, their goal was to get both firms to pump cash into Google while leaving neither in charge of the company. While the venture firms battled with each other for dominance, control, and the right to be the sole investor, Brin and Page would shape Google's destiny and remain the majority owners. If not, the founders would have to find another source of money. So what if that was not the way people were used to doing business? They would do it their way, or no way at all. There would be no compromise. It was as simple as that.

In the frenzy of the dot-com boom, both John Doerr of Kleiner Perkins and Michael Moritz of Sequoia Capital had grown tired of watching an endless stream of PowerPoint presentations about new business ideas. It was hard enough to figure out which entrepreneurs to bet on; when ideas were flashed on a screen, they were also forced to guess which of the dozens of new concepts and technologies they were seeing would catch on. For these two titans of Silicon Valley finance, Sergey Brin and Larry Page were a breath of fresh air. Instead of a PowerPoint presentation, they came marching in with a working search engine technology that was superior to anything that Doerr or Moritz had ever seen. The Google Guys seemed super bright and extremely brash. They knew it all. But they also had connections, Stanford pedigrees, fire in their bellies, and focus in their beings. Unlike ordinary people with good ideas who might not be able to see them through, it seemed clear that these guys would do whatever it took to execute.

The biggest questions revolved around how to value a search technology that had no real business model associated with it, and whether any serious investor would ever be able to work with the two of them, given that they wanted as much money as possible while retaining as much power as possible. Neither Doerr nor Moritz had a particular problem with the long-term strategy that Brin and Page wanted to employ. Both their firms were making plenty of money elsewhere, cashing in on the dot-com craze. Stanford professor David Cheriton, a confidant of the venture capitalists, vouched for the Google Guys' honesty and work ethic when he helped set up the meetings. And there was something else. Despite the deadly serious way in which the guys approached their PageRank technology and computer hardware strategy, they came across as genuinely nice people who could converse about a wide array of subjects. They were atypical. And that made them desirable.

Moritz had first met Brin and Page through Yahoo's David Filo back when they were still students at Stanford. Moritz and his firm, Sequoia Capital, had backed Yahoo with $2 million and reaped a big gain from Yahoo's $32 million IPO in 1996. Around that same time, Brin and Page were gathering information about starting a company, including valuation methods and other techniques, so they could work out an agreement with Stanford that would facilitate the patenting of PageRank and enable them to license it from the school. They also had been advised that meeting some venture capitalists early, before they needed money, was a good idea. "They came by the office and they were curious about how to start a company. But it was fairly quick, one of those conversations of which we have many," Moritz recalled. "I didn't think much about the conversation afterward."

As 1999 wore on and Google began to run short of cash, one of its angel investors, a Silicon Valley money manager named Ron Conway, contacted Moritz and asked him to meet with Brin and Page.

"Ron Conway pinged us about them," Moritz said. "We also heard about them through the Yahoo guys as well. This was the spring of 1999, so everything was done in a pell-mell rush. Those were fairly frenetic circumstances and times."

Moritz remembers being very impressed with the demonstration of the Google search engine. "We had several meetings at Sequoia and down in Palo Alto where they had a little office on the main drag," he said. "They had the Google beta site running, so it was reasonably easy to detect the quality of the difference between the search results they were providing and what others were providing. Their original business idea had nothing to do with advertising. It was aimed at licensing the technology to a variety of other Internet companies and enterprises."

How did Moritz evaluate their potential? His answer offers clues as to how one of Silicon Valley's most successful venture capitalists makes decisions: it is more art than science. And his being an investor in Yahoo also played a role, illustrating how multiple sets of relationships can influence decision-making that might otherwise appear linear.

"We have been wrong on lots of occasions," Moritz explained. "It was the quality of the service that Google provided that was demonstrably better than you could get anywhere else. That is why we invested. And as the Internet developed, we thought search would be more important, not less important. These two, Larry and Sergey, were very smart people. We had some benefit from being around the evolution of the Internet. The other thing that happened was that Yahoo had had licensing relationships with a considerable number of search vendors. They had relationships with Open Text, with AltaVista, and with Inktomi. And Google was the latest to come along.

"The Yahoo people were very interested in Google as a search engine to power their service," he said. "And they also were interested in wanting to help us become an investor in the company because they thought it would help Yahoo." Moritz said Sequoia was inclined to invest in Google in part to assist Yahoo—"for us to help ensure that Yahoo was taken care of. They were viewing Google correctly. Nobody understood in 1999 how things were going to evolve. Google was a potential vendor to Yahoo. It seemed to us that the Internet had spawned two useful applications: one was email and the second was search. They had built a better search trap."

Moritz said he also recognized that Brin and Page, together, had the right stuff. He had seen over and over again how start-up companies founded by pairs of entrepreneurs who shared a common vision had a greater chance for success than lone individuals. It had happened at Microsoft with Bill Gates and Paul Allen. It had happened at Apple with Steve Jobs and Steve Wozniak. It had happened at Yahoo. And maybe, just maybe, it could happen at Google. "They were a pair of unusually smart men. That was very evident. In our business we meet lots and lots of people and over time you develop a sense of who the special individuals are partly because of what they have done or are doing, and partly because of the way they express themselves. They had a great sense of purpose, which is a prerequisite for anyone who is nutty enough to want to start a company. That burning sense of conviction is what you need to overcome the inevitable obstacles."

While Moritz focused on Brin, Page, and the search engine they had built, John Doerr of Kleiner Perkins liked the long-term dynamics of the Internet and the potential and promise that Google represented in that evolution. More than most venture capitalists, Doerr was willing to bet big on the technologies that would benefit from increasing Internet use in smart ways over many years. During the dot-com boom, he eschewed the conventional wisdom that things were getting overheated and overhyped, and insisted that the Internet's potential was much greater than people realized. Doerr was the venture capitalist who had made a fortune backing Compaq Computer, Sun Microsystems, and Amazon.com before most people understood the concepts driving them. In fact, he already had a connection to Google through Amazon founder Jeff Bezos, who was both an early investor and an informal advisor to Brin and Page. Doerr's experience and status as the star venture capitalist in Silicon Valley wasn't lost on the guys; they knew his involvement with Google could help transform their vision and ideas into solid business deals. Doerr had also been an early investor in America Online, the biggest Internet provider—and a giant potential customer for Google. Larry and Sergey valued the money such deals would bring. It was the essential ingredient they needed if they were to realize their dream of building the best and most comprehensive search engine on the planet. An investment from Kleiner Perkins, led by Doerr, would put the imprimatur of success on their start-up.

In the spring of 1999, Moritz and Doerr both decided that they wanted their respective firms to invest in Google. Brin and Page found themselves exactly where they wanted to be, but with a vexing problem: the two venture capitalists refused to invest jointly in the company, so Google risked losing them both. Each firm wanted to call the shots and lay claim to Google as "its" deal. Each was too big, in its own right, to cede control to the other—and neither needed the deal if it meant being a minority partner. That wouldn't work, not in the pecking order of Silicon Valley. Both Kleiner Perkins and Sequoia controlled too much money to bother sharing investments in start-ups. That wasn't how you hit home runs in the venture capital business, at least not at the most prestigious firms on Sand Hill Road.

It was a dilemma for Brin and Page. On one hand, they needed money fast and had two firms offering it. On the other, maybe they could raise cash without giving up control if they could persuade both firms to invest. It was a big "if," but that was the route they decided to pursue, even it meant losing them. Fortunately for Google, two of their early angel investors, Ron Conway and Ram Shriram, had relationships with Moritz and Doerr respectively, and sought to resolve the unusual standoff. As the weeks dragged on, the guys could sense why venture capitalists had earned the moniker "vulture capitalists," and they began to think Google might be better off without either of them.

Brin and Page asked Conway, who had other contacts as a money manager, if he could assemble a group of angel investors instead. An array of passive investors meant Larry and Sergey would remain in charge, and they told Conway that was a route they were prepared to go, adding that time was of the essence since they were running low on cash.

Instead of reaching out to other potential angels, however, Conway contacted Shriram. They decided to tell Moritz and Doerr that unless they could find a way to work together, the Google Guys were going to walk, and they weren't bluffing.

Although this was at a time when legions of entrepreneurs were stopping by Kleiner Perkins and Sequoia to seek funding for new dot-coms, alarm bells went off: there was something especially promising about this duo. The venture firms put aside their own egos, and within days, Conway and Shriram had the deal worked out. Kleiner Perkins and Sequoia Capital would each invest $12.5 million in Google, splitting the $25 million venture portion of the deal in half, and accede to the demand that Larry and Sergey remain in charge with majority control. But with that much money on the line, Doerr and Moritz attached one condition to the funding: a promise that the entrepreneurs would hire an experienced industry executive to help them transform their search engine into a profitable business. It was a rational request given that the company didn't even have a business plan. And Brin and Page agreed to it, sensing that if they had the $25 million in hand, and voting control, they would have the leverage and could push off any decision to hire somebody as a chief executive well into the future. One thing was certain: they had no intention of hiring anyone whom *they* would report to.

On June 7, 1999, less than one year after they took leave from Stanford, Brin and Page issued a press release announcing that Kleiner Perkins and Sequoia Capital had agreed to invest $25 million in Google Inc. On Stanford's campus and around Palo Alto, jaws dropped. It was a huge sum. These two venture capital firms didn't coinvest, yet Doerr and Moritz were both joining Google's board of directors. And somehow, the two guys, who had always seemed a bit too confident to some of their classmates, had managed to extract an enormous amount of money seemingly without giving up anything in return. All indications were that the Google Guys had a dream deal: the money they needed to build their beloved search engine, and the control and power they needed to call the shots.

"We are delighted to have venture capitalists of this caliber help us build the company," Page said in a formal statement. "We plan to aggressively grow the company and the technology so we can continue to provide the best search experience on the Web."

Brin added an appropriately grandiose comment, his confidence higher than ever. "A perfect search engine will process and understand all the information in the world," he said. "That is where Google is headed."

The press release described their patent-pending PageRank system for ranking search engine results as consisting of 500 million variables and two billion terms. This led, it said, to unprecedented accuracy and quality that extended Stanford University research into large-scale data mining of the Web. In fact, if the press release had not contained statements from Doerr and Moritz, too, some on Stanford's campus would have thought it was a spoof.

"Google should become the gold standard for search on the Internet," said Sequoia Capital's Moritz. "Larry and Sergey's company has the power to turn Internet users everywhere into devoted and life-long Googlers."

Doerr added, "Search is extremely challenging, and improvements in the technology are significant. One hundred million Web searches are performed every day. Quickly finding the right information is critical for Web users in many professions. Google revolutionizes search technology and delivers information in a way that focuses on the user."

The announcement included details about the funding as well as additional information about Google, its impressive list of investors, and its growth rate of 50 percent per month. All of this put the company in the global limelight, giving it the opportunity to grow further through free media publicity. The following day, Sergey and Larry sent an email to friends: "This was an exciting month for us as we secured funding so that we can continue to improve Google in new and exciting ways. Our capacity is still going up (thanks to you!), and we've been expanding to meet the demand. This month we've put in even more servers to ensure a faster user experience (we've started ordering our computers in 80 packs, up from our previous increment of 21 packs). We have also been working to make sure the duplicates are removed from search results and we are working on some new features (sshhh!) that we hope will improve our users' search experience."

It was a heady moment for Google and its founders. Still, the press release and news coverage of the deal, replete with accolades about the search engine and bold statements about the future, failed to answer a central question. The mystery remained: how did Google plan on making money?

CHAPTER 6
Burning Man

In late August 1999, Larry and Sergey, along with other Google employees and friends, piled into cars and set off for a desolate stretch of Nevada's Black Rock Desert. They untethered themselves from the Internet and would have no cell phone reception out on the playa. Nor would they have anywhere to purchase water or food, so they carried a week's worth of supplies with them. They were not alone: a motley assortment of other technologists, artists, anarchists, intellectuals, and free spirits were also making the trek—some 18,000 in all. Undeterred by massive traffic jams on the narrow roads north of Reno, or an unforgiving climate where temperatures range from 40°F to 110°F, they were bound for a temporary city not found on maps and inhabited only briefly each year before disappearing again. Over the years a festival of bizarre art and bacchanalia had sprouted there, and Sergey and Larry found it too much fun to miss, whatever the demands of their new business. Besides, some of Google's most loyal users and potential partners would be there too. Attracting them all to this starkly beautiful setting was a magnet with a poignant moniker.

Burning Man.

Its name alone conjured up images of fiery passion and pagan ritual. Rising 40 feet and illuminating the western sky in neon, a giant wooden figure known as "the Man" exacted a curious spiritual grip on Larry, Sergey, and their comrades in the desert. There was no single interpretation of what the effigy stood for. The same was true for the festival as a whole, making Burning Man both a social gathering and a highly personal sojourn.

Pilgrimages to this pulsating mélange of dust, flesh, and heat had become a rite of late summer throughout the Bay Area, and especially in Silicon Valley. Everyone from serious-minded engineers to hard-core partyers was fleeing the ordinary limits of society to return to this "temporary autonomous zone" of freedom in the desert. Friends and lovers arrived together, or discovered one another on the vast expanse of the playa, where nudity and drugs mingled easily. The makeshift streets that the Google Guys traversed at Burning Man were named after planets in the solar system, with the cross streets identified by times on a giant sundial etched in the desert. Filling out the radial city were trucks, tents, RVs, and outlandish themed campsites. And at the center stood the Man, at once otherworldly and oddly captivating against the nearby mountains.

Burning Man went largely unpoliced, but it had certain principles. Buying and selling were forbidden, except at a central café that dispensed coffee and ice. Leaving the desert environment unspoiled was also fundamental. "It's our land-use ethic that distinguishes us from Woodstock," said Harley K. Bierman, the festival's earth-guardian organizer. "Burning Man is a social experiment where an entire city gets built, lived in and then disappears. And that is the biggest performance art piece of the whole event—the art of leaving without leaving anything behind."

Before Larry and Sergey departed for Burning Man, they created some experimental art on their search engine's homepage. Seeing an opportunity to infuse new life into the logo, they incorporated a rendering of the Man in the second *o* of Google. To an outsider, the stickman logo looked crude, hastily assembled. But to those in the know, it signaled where the Google crew would be that week.

The impulse to tinker with the design triggered an organic change. Heading out the door to a festival celebrating ingenuity, they unknowingly had given birth to the Google doodle. For the Googlers, dressing up the logo that first time filled a void. With the company's founders and friends all heading to the desert, it was a way to communicate to users that if the search engine went down, nobody would be around to fix it. Marissa Mayer, an engineer who joined Google that summer, recalled that the stickman was "more of an out-of-office notification than anything else—it said, 'We're all at Burning Man.'"

Burning Man had its small, accidental beginning in 1986 as a celebration of the summer solstice on a San Francisco beach, when a group of friends burned an eight-foot wooden effigy of a man. By the time Larry and Sergey moved to northern California in the 1990s and began attending, it had transformed into a countercultural mecca, drawing thousands to a temporary oasis "liberated from nearly every context of ordinary life," in the words of its founder, Larry Harvey.

Walking the playa at Burning Man, Brin and Page drew inspiration from a series of giant art installations, commissioned in advance, that reflected a blend of creativity and engineering, talent and technology. The event's theme for 1999 was the "Wheel of Time," and various spaces surrounding the Man were designated for the art, based on whether it expressed something about the past, present, or future. With the new millennium fast approaching, this was a chance to be exposed to a cornucopia of fresh ideas and bond with others. Unlike some artists and engineers, who constructed elaborate projects over the course of weeks or months, Brin and Page preferred the spontaneity of the experience. "Less planning is better," Page said.

Larry liked to roam the festival with a camera, capturing its images. The sensory overload made it impossible to take in all at once. (A number of Burning Man photos, including one with a sweeping panoramic view, would later hang on the walls of Google's headquarters.) The exhibit "Sands of Time" offered Larry and Sergey a self-guided tour around the sundial, with prerecorded voices of famous scholars talking about the concept of time. Another, the eerie, three-story-tall "Bone Tree," traveled around the wheel of time and was sculpted entirely from weather-bleached bones. Around the Man they found "L2K," a circle of 2,000 flashing lights buried in the ground and wired into eight golf-cart batteries. That display, like Burning Man, looked its best at night, when fire, luminous artwork, dancing, and music of all sorts transformed the scene and encouraged exploration.

For many people, the Burning Man experience includes heavy drinking and drugs. Not Larry and Sergey. Instead, they found the energy and ingenuity intoxicating, and relished the numerous chance encounters with clowns, exotic dancers, or old friends. "It is a wonderful chance to experiment with different ways of thinking and interacting with people," said Brad Templeton, chairman of the Electronic Frontier Foundation, who has spent time with Brin and Page at the festival. "You go there to be creative and appreciate other people's creations. You take a vacation not just from regular life but from the way the world works."

Some Silicon Valley colleagues, clad in costumes or made over with body paint, were not immediately recognizable to Larry and Sergey. "There is a lot of silliness and a lot of whimsy," said Tamara Munzner, their Stanford classmate. Brin and Page could crawl through a 30-foot wind socket called the "Black Hole," smile at an orange elephant riding around on a bicycle, or view a kinetic sculpture that created the illusion of a person running while on fire. "They were drawn to it because of the technical and creative wildness," said Sean Anderson, another Stanford classmate.

Brin and Page harbored strong beliefs about what made for productive culture, community, and ethics in their young company, and the rigidly noncommercial nature of Burning Man, where advertising is banned, gave them perspective. The festival "values creativity and doesn't value money at all," said Stewart Brand, a futurist and friend of the pair. "There is nothing a billion dollars can buy them at Burning Man except a cup of coffee."

The harsh physical environment challenged Larry, Sergey, and other participants. They had to rely on each other to survive. By encouraging sharing and teamwork, the atmosphere at Burning Man contained elements of the culture they were creating at Google. Sergey and Larry also liked Burning Man's "participants only" philosophy, and the way it encouraged people to push boundaries. "Both Sergey and Larry get a lot of their inspiration from Burning Man," Brand said. He added that they liked to "wander and sleep where tiredness finds them."

On the last night, in a stunning display of fire and pyrotechnics, the Man was set ablaze as the festival reached its climax. Just before the ritual burn, a group including Googlers, Templeton, and others staged a mock protest. They marched around chanting, *"Don't Burn the Man! Don't Burn the Man!"* Then, after the effigy was torched, they threw their signs into the fire, joining thousands in a massive celebration that encircled the raging bonfire.

"The burning of the Man is spectacular," Templeton said. "After the Man collapses, everyone rushes in and is dancing and singing—and some people take their clothes off if it is not too cold. You have an experience you won't find in the outside world."

Two months after Burning Man, Marissa Mayer was in the office late on a Saturday night, Halloween eve, trying to wrap up some work at Google before a long vacation to Europe. She was still writing emails at two in the morning when Sergey called out, "Marissa, look at this, look at this!" Brin, with the spirit of Burning Man still aglow, had designed a Halloween-themed logo on his computer consisting of two orange pumpkins placed over the Google *o*'s.

Mayer was not impressed. She thought it looked terrible, "clip arty."

"Put it up on the site," Sergey said.

"You want me to put *this* on the site?" Mayer said. "Do you see that I can see the *o*?" One of the pumpkins wasn't properly centered, making some of the red letter below it visible. Brin responded, "We are all here, excited for Halloween. We should show people in the world that people at Google care about Halloween."

Google users, it turned out, loved the clip-arty pumpkins. The logo drew a huge response and even scored a mention on Slashdot, the No. 1 must-read techie Web site.

Sergey had other ideas for dressing up the logo every now and then, to delight and surprise. He wanted to fete other holidays and do longer, evolving stories, such as a plant that would grow over the course of a week. Someone talked Brin out of that one, but the Google doodles were a hit with users, who wanted more. Soon it fell to a committee of Mayer and two others to decide which holidays and events to grace with a doodle. In the early going, they paid tribute to several independence days, but these always made someone unhappy. Religious holidays would do the same, they figured, so they steered toward politically correct holidays and events like Chinese New Year, Earth Day, and the Olympics, posting usually one or two doodles per month.

In November 2001, Mayer pegged Dwali, the Indian festival of light, for a doodle. Logos that had worldwide appeal were especially fun to do. She asked whether Dwali was religious, and was assured it was not. But on November 13, the day before it was to run, a colleague informed Mayer that Dwali was based in Hinduism and they probably should pull the plug. Disappointed, Mayer googled an online holiday calendar and discovered that the next day was Claude Monet's birthday. Artists had global appeal, she reasoned, so they certainly could run a Monet doodle worldwide. Besides, her mom was an art teacher.

Mayer called Dennis Hwang, a computer science grad with a studio art background who had become Google's resident doodler. Hwang drew his first doodles during a summer internship at Google in 2000; after joining the company full-time, he began devoting several hours each week to refining his doodlecraft. But Hwang was out sick that day, so Mayer decided to give it a go herself, using the graphics program GIMP that had a built-in

"Impressionism" filter. "Looked like Monet," Mayer said of her creation. "That got Dennis scared we were going to run it Gimpified." So, with a 103-degree fever, Hwang spent 20 minutes that night doing his own Monet from home.

The Monet logo, still one of Hwang's favorites, was another success, and began a tradition of commemorating artists' birthdays with doodles. The portfolio has since expanded to include scientists, famous discoveries, even the occasional entertainer. Not all were greeted kindly: the estate of Salvador Dali made Google take down a Dali doodle a few hours into the run. But overwhelmingly, the doodles evolved into a treat for users around the world, some of whom came back to the site day after day just to see if there was a new one. "Logos touch people," Mayer said.

Inevitably, doodles also generated their share of amusing emails. There were always a few computer users who saw a doodle and thought Google had permanently changed its logo. When Hwang did a tribute for Michelangelo's birthday, substituting the famous *David* sculpture for the letter *l* and making the entire logo appear to be chiseled from rock, one user clearly missed the reference and sent in a blunt appraisal: "Rockman sucks."

Around the time that Marissa Mayer was hired in 1999, Google was also looking to recruit an engineer to analyze, test, and improve the layout of the search engine's Web site. Though this process consisted of seemingly inconsequential things like tweaking text sizes and fonts, it was critical to how users perceived the Web site and whether they would return to it.

Reviewing the résumés, Mayer's boss realized that she was the best qualified. A native of Wisconsin, Mayer had a master's degree in computer science from Stanford and had also studied linguistics and psychology. She was issued strict instructions on how to tackle this the Google way. "Don't bring any new opinions to the table," her boss told her. "You are not allowed to have an opinion. Your job is to get data."

Google's Web design at the time was as plain and simple as its behind-the-scenes architecture was complex. First-time users were often surprised by the barebones look. In stark contrast to the cluttered sites then in vogue, Google stayed clean and spartan. The "less is more" approach made perfect sense: Larry and Sergey had put a premium on speed in every aspect of building their search engine, and a graphics-heavy homepage would slow it down.

Users appreciated the speed of Google, and those who spent a lot of time on the site began to form a strong connection to it. Some became so familiar with its look that the subtlest of changes caught their eye. Mayer quickly learned just how sharp, even obsessive, their users were. One vigilante sent Google an anonymous email every so often just listing a number, like 37 or 43. Eventually Mayer and her colleagues figured out it referred to the number of words on the Google homepage—the implication being that someone was keeping track, so don't screw up the design.

In December, Mayer rolled out one of her first big changes to Google's design: a new font. Following orders to gather data, not opinions, she had researched different fonts for their legibility on a computer screen, and decided on Verdana, a sans serif font. At the time, Google was using a serif font, but Mayer found that fonts without serifs (the tiny curlycues on certain letters) made skimming search results easier. Thinking nothing more of it, Mayer left the office for a half-day outing with Google's other female engineers to have tea at a fancy hotel in San Francisco. She returned several hours later to several strongly worded threats. Verdana was inappropriately sized! It was off by two point sizes! How could Google lose serifs and become sans serif?! Whoever ruined Google . . . ! The intensity of the reaction surprised Mayer, but she learned a valuable lesson for future design changes: you need data, but you also need people to test it.

The next month, Google invited 16 people to the basement of the Gates Building at Stanford, where a team of four from the company observed their behavior. "We put two people at each computer, so they'd talk to each other instead of talking to us," Mayer explained. The testers were told to use Google to find the answer to a trivia question: Which country won the most gold medals in the 1994 Olympics? They typed www.google.com, watched the homepage come up on the screen, and then they waited. Fifteen seconds went by . . . twenty seconds . . . forty-five. Mayer wondered what was going on, but didn't want to interfere. Finally, she asked them, What are you waiting for? The rest of the page to load, they answered. The same thing kept happening all day, Mayer recalled. "The Web was so full of things that moved and flashed and blinked and made you punch the monkey that they were waiting for the rest of it to show up." From that inauspicious start, Mayer's team concluded they had to beef up the copyright notice and footers at the bottom of the page, not for legal reasons but to let users know "That is it, that is everything. Please start searching."

Mayer's team learned a lot that day about other ways to improve Google's homepage. One tester wondered aloud whether Google was a legitimate company because the Web site looked so unpolished. After Mayer told her the firm had plenty of employees, "she asked if I was the psychology department in disguise making up a fake company called Google."

When Google went looking for someone to ramp up its computer network, Larry and Sergey hired a brain surgeon, Dr. Jim Reese, who had degrees from Harvard and Yale Medical School. He had been working at a Stanford research lab before he joined Google in 1999 as employee No. 18. Named Google's operations chief, Reese managed the company's burgeoning collection of computer hardware.

In the fall of 1999, Google went on a shopping binge. Flush with cash and encouragement from the two venture capital firms that invested in Google, Sergey and Larry had the resources they needed to grow the company aggressively. For Google to reach a position where it could bring in substantial revenue, the company needed to buy additional computer parts and memory so it could expand its network.

They no longer needed to scour the Stanford loading docks for computers. They had graduated to the next level: hopping in the car and driving to Fry's, a giant Silicon Valley electronics emporium. There, they stocked up on garden-variety PCs, disks, and memory. Back at the Googleplex, they ripped apart the machines, disposing of all the unnecessary parts that would eat up computing power and resources. Then they built streamlined computers, stringing them together with software, wiring, and the special sauce that made Google lightning fast.

"We want to get the most computational power per dollar that we can," said Jeffrey Dean, one of several engineers Google plucked that year from the lab that had spawned the AltaVista search engine.

Dean and other Googlers from this era love to tell the story of how they cobbled together a virtual supercomputer from cheap, commodity PCs. Rather than spending $800,000 on a high-end system from IBM, he said, they went to RackSaver.com, where for just $250,000 they found a rack of 88 computers that provided comparable processing power and several times more disk storage. They also used the free operating system Linux rather than buying software from Microsoft. Those savings gave Google a significant edge over competitors, even those able to match them dollar for dollar in capital spending. For every dollar spent, Google had three times more computing power than its competitors.

Because Google's basic PCs did not have the built-in safeguards and redundancies of the elaborate IBM supercomputers, they were more prone to failure. Just like the ordinary desktop machines they resembled, Google's PCs would last an average of two to three years before needing to be replaced. Even healthy machines might be retired after a couple of years, since by then they would be slower than newer machines. Given the number of computers Google was using, several were destined to fail each day. Sergey and Larry elected to deal with the constant failures through software, bypassing any machines that died rather than manually removing and replacing them.

Enter Dr. Reese, who instead of performing operations on humans would assist in devising software so that Google would be fast and reliable, no matter what occurred. By spreading data and computing tasks over a large number of computers in multiple locations, he and his team created a system that could handle some miscues without crashing. With this key element of Googleware in place, Dr. Reese the brain surgeon could monitor the entire precious network from a single location, instead of having to constantly patrol each of Google's off-site data centers.

Most of Google's computers were housed in these giant centers—nondescript, temperature-controlled warehouses that leased space to companies that wanted to keep their systems online and secure. During the hotly competitive late '90s, data centers charged companies by the square foot rather than for the electricity they consumed, giving Google an incentive to cram as many machines as possible on top of each other. This was a money-saver for Reese and his team, but as electricity costs soared, some of the data centers went bankrupt and Google was forced to lug its computers elsewhere. After doing this once or twice, they made sure their computer racks were outfitted with wheels.

In the buying spree, Google expanded from having about 300 computers when Reese joined to 2,000 one month later, and twice that number by the following summer. With this sort of expansion, it was imperative that Google build redundancy into the system, so that if one cluster of computers went down—or was completely lost in an earthquake—other clusters would have stored copies of the Internet and other data to pick up the slack. At the time, Google had two data centers in northern California and a third in the Washington DC area, later adding many more across the U.S. and overseas.

The value of having multiple copies of everything became clear when a fire broke out in one of Google's data centers. Larry and

Sergey weren't at Burning Man anymore. This was the real deal. As six fire trucks doused the blaze, Google's redundant systems took over, enabling it to continue delivering rapid results. Google had proven itself reliable, and tens of thousands of information-hungry computer users had no idea that anything had even gone wrong.

CHAPTER 7

The Danny Sullivan Show

Danny Sullivan, a 30-year-old newspaper reporter for the *Los Angeles Times* and *The Orange County Register*, was one of the first to catch Google fever in the late 1990s. When he quit the paper-and-ink world of newspapers and went off to work with a friend who was designing Web sites, he was hardly alone. In California, where Sullivan had been born and spent virtually his entire life, everyone, it seemed, was talking about the Internet, investing in a dot-com, or going to work for a start-up. "It was big, and I wanted to be a part of it," Sullivan said. Ironically, the turning point for him came one day, not long after he had made the leap, when an angry customer whose Web site listed available jobs in Orange County became furious with Sullivan and his business partner because the customer's Internet site was not showing up prominently enough in search results. The rage caught Sullivan off-guard. He was a friendly and analytical sort, and he wasn't used to being chewed out. Moreover, he couldn't understand what had gone wrong.

The customer's fury served as a catalyst for Sullivan, sending him on a mission to discover how search engines actually gathered data and generated rankings and results. For the next several months, Sullivan poked around the Web, asked a lot of questions, and grew increasingly fascinated by the hidden world of search engines. Among other things, he discovered that the automated crawlers sent out by AltaVista and its top competitors to scour the Web randomly omitted certain information from their databases. It wasn't completely clear why, though it seemed likely that they were simply not able to keep pace with the growth of the Web. Sullivan now realized that he was onto something big; he had found a subject that was extremely important for Web site owners to understand if they wanted to build traffic and garner attention. He published a study, "A Webmaster's Guide to Search Engines," on the Internet, and explained that search engine marketing was tricky business. Even though there were numerous search engines, they remained primitive in many ways, copying the Web slowly and routinely leaving out important data. While it was a good idea for site owners to try to improve their rankings on all the engines, there were too many of them, coming at the problem in too many different ways, to connect with them all. So instead of a mass market approach, Sullivan suggested in his study that Web publishers focus on AltaVista, Excite, and a few other leading engines. It was sound advice.

By the time Sullivan's report was online, however, the company he had gone to work for had gone bust. Nevertheless, Sullivan was in the right place at the right time. He received a lot of positive responses to his study and went into business himself, becoming an Internet consultant with a Web site he named Califia.com.

Convinced that the Internet enabled him to consult from anywhere, he and his wife, a native of Great Britain, moved from California to London in 1997 so they could be closer to her family. He also began publishing a regular online newsletter called Search Engine Watch. Based on the strong reaction to his initial study and the dialogue sparked by occasional updates to his Web site, Sullivan sensed that the Internet was a place where a single individual with expertise in a particular area could make a living by publishing a newsletter online. Unlike traditional newsletters, his online publication had no cost of distribution, and he could instantly post up-to-the-minute news or insights about the evolving search engine industry. Its popularity grew quickly. In November of that year, Sullivan sold Search Engine Watch to JupiterMedia, a leading online market research and news network, but he retained the Web site's name and his identity as the guru of search. And he soon added the role of onstage emcee to his portfolio as he toured several major cities hosting annual Search Engine Strategies conferences.

Sullivan knew it was important for him to visit Silicon Valley from time to time so he could stay in touch with fresh developments there and meet the new players emerging in search. Aware of Google when it launched in 1998, he wondered whether the Stanford-born search technology, which was quite impressive, would turn into a viable business with Brin and Page at the helm. "They didn't seem worried about how to make money," he recalled.

Google first showed up on the radar screens of those beyond its circle of devotees in 1999, when it received the $25 million in venture capital funding. But for quite a long time, even after its brand became relatively well-known and the search engine was cranking out millions of responses for free to computer users around the world, the company struggled to generate cash. With the notable exception of two companies, Red Hat and Netscape, nobody was willing to pay for the right to license the Google search engine. In that first year after Kleiner Perkins and Sequoia Capital invested in the firm, it looked as though the skeptics who said they had wasted a huge sum of money might be proven correct. When Sergey Brin roller-skated across the stage at the 1999 Search Engine Strategies conference hosted by Danny Sullivan in San Francisco, people laughed and took notice, but Google's investors were not impressed.

"The original business idea was aimed at licensing the underlying search engine technology to a variety of other Internet companies and enterprises," said Sequoia's Michael Moritz. "During the first year we collectively had concern that the market we were pursuing was more difficult and more intractable than we had originally anticipated.

"The conversations with potential customers and negotiations with potential customers were protracted," Moritz went on. "There was a fair amount of competition, and we didn't have a direct sales force. The customers were very harsh on the prices they were prepared to pay. It was clear if we were going to pursue that path, it was going to be a brutal path."

Brin and Page were not discouraged. They knew they had a better search engine and they sought to forge a relationship with Danny Sullivan so he could help them spread the word globally without their having to spend money on advertising. It worked. Positive mentions on Sullivan's Web site, and the emails that followed, served as one of the important ways Google marketed itself without advertising. It was a symbiotic relationship that would work for the entrepreneurs, for Google, and for Danny Sullivan too.

Although Google was averaging seven million searches per day by the end of 1999, its revenue from licensing deals remained small. While Brin and Page didn't care about getting rich, they didn't want Google to flounder. If the business could not sustain itself, they would not be able to fulfill their vision of making all of the world's information easily available to users without charge.

The two of them wondered how best to navigate this maze. Anything that compromised the relationship of trust they had with their users would be unacceptable to them, even if it generated a lot of money for Google. They felt conflicted about advertising, and had included a strongly worded statement on the subject in a 1998 paper they published about Google technology. Ad-funded search engines are "inherently biased towards the advertisers," they wrote, suggesting that "the better the search engine is, the fewer advertisements will be needed for the consumer to find what they want." At the same time, advertising was a type of information that some users might want. Perhaps not all ads would violate the Don't Be Evil motto that Brin and Page had adopted for their company.

Yahoo, the giant Internet site, was dependent on ads. It saw broad-based search as a sideline feature that it contracted out to others. Yahoo's main business was being a home for millions of registered email users and providing them with excellent service, content, and community. When it came to search, Yahoo's manually edited, categorized directories, which harkened back to the original vision of its founders, couldn't keep up with the mushrooming Web, and the directories became less valuable. To address that problem, Yahoo relied on other firms, such as Inktomi, to crawl the Web as an unbranded search engine and augment the results Yahoo served up to users. Microsoft, for its part, didn't see consumer search as a major business opportunity and invested in numerous other areas. Both Microsoft's MSN division and America Online saw email as the golden goose that created an audience for advertisers, and they viewed search as considerably less important in the mix. They, too, were relying on third parties to provide them with unbranded search results.

All the while, Danny Sullivan watched carefully as Larry and Sergey seized the opportunity to hire bright technologists and focused on doing one thing: search. Google's revenue might not be growing rapidly, but its employee brainpower was. Instead of paying these engineers big salaries, Google offered them mediocre pay and thousands of stock options that might prove valuable someday if the company prospered. This was the typical Silicon Valley bet that bright engineers made. Sergey and Larry, who interviewed every prospective employee themselves, had their pick of talent, since they were hiring while everyone else was firing. It also helped that Google was a private company during the technology bust. For all the pressure mounting on Google because of its lack of revenue, it was nothing compared to the angst at publicly held companies that had sold stock to investors at high prices only to find that having ".com" in their name was not enough to guarantee survival.

With a superb search technology and a growing edge in the quality of its employees, Brin and Page sought to figure out how to keep Google afloat without compromising their lofty principles.

Rather than focus on licensing its search technology to businesses, they decided to concentrate on profiting by allowing advertisers to reach their growing and loyal legion of users. Google would continue to keep search results free—just as television networks offered entertainment and news for free—and would look to make money by selling unobtrusive, targeted advertising to businesses on the results pages.

Sullivan, for his part, did not view all advertising as inherently evil, but he did feel strongly that the free search results delivered by a search engine should not be influenced by advertising or other payments in any way. And he therefore had serious questions about any search engine that sold ads or accepted payments in return for guarantees that it would crawl and include a Web site in its free results. He let readers of Search Engine Watch know that he found that approach problematic. A few search engines, hungry for cash, accepted money from Web sites in exchange for placing the sites in search results. Brin and Page saw this as a "particularly insidious" form of bias and eschewed any such payments. They also disliked the flashy, irrelevant banner ads that littered the Internet. But there was another path to profit: running targeted "text only" ads that were triggered by users' specific search requests.

"We are about money and profits," Brin said, revealing his conversion from strident student to pragmatic company president. "Banners are not working and click-through rates are falling. I think highly focused ads are the answer."

One company that caught Brin's attention, for the simple reason that it seemed to be making money by selling ads to accompany search results, was GoTo.com, later renamed Overture Inc. Although most consumers had never heard of Overture, it provided the ads that appeared with search results on Yahoo, America Online, EarthLink, and other major Internet sites. In fact, in contrast to the backlash against annoying pop-up and banner ads, search-related advertising was just about the most appealing game in town, and one that was growing. Brin and Page began studying

Overture and immediately found aspects of its approach distasteful. Among other things, it would sell guarantees that Web sites would be included more frequently in Web crawls, provided a business was willing to pay extra for it. The Google Guys had two choices: they could hire Overture to sell ads to display alongside search results on Google.com, or they could try to sell advertising themselves.

The decision was not difficult. Google wrote all its own software and built all its own computers. If Sergey and Larry believed in anything, it was their own ability to get things done. If they could conquer the sophisticated venture capitalists on Sand Hill Road, why couldn't they sell ads themselves and keep 100 cents of every dollar, rather than share the proceeds with Overture? Doing it themselves would also give them total control, enabling them to avoid potentially dangerous conflicts and any perceptions that might damage Google's stature as a trusted brand name. So the two decided to try to emulate aspects of what Overture had started in 1998 on the ad side, with some tweaks. The newly emerging business strategy was simple: continue to produce free search results, and profit by selling ads. The key was to make it clear they wouldn't bias the search results.

"The thing that always came across loud and clear was their dedication and seriousness about wanting to do the right thing," Sullivan said.

Brin and Page talked to many people before embarking on the path toward ads, to ensure that they didn't make any mistakes. They became persuaded that, just as there was a clear distinction between news stories and ads in newspapers, they could achieve the same thing on Google.com. But they hated to clutter the clean interface that had been its calling card from the start, so they kept the homepage free of ads, and they developed strict standards for the size and type of ads they would display elsewhere. They decided also to have a bright line on the results page that separated the free search results from the ads, which they would label "Sponsored Links." That way, nobody could argue that the search results were combined with the ads, yet the ads would be clicked on more often under the heading "Sponsored Links" than if they were simply labeled "Ads."

At the start, Google priced its ads the way traditional media companies did, based on the size of the audience. After talking it over with various experts, and testing different screen layouts, they decided to display the ads in a clearly marked box above the free search results. They wanted the user to have a good experience, so there would be no pop-ups or other graphics that would interfere with a Google search. The ads were to be brief and look identical—just a headline, a link, and a short, haiku-like description. Initially they were sold one by one, mostly to larger businesses that could afford hefty ad campaigns. But by utilizing their own technology, they soon moved to a model that enabled advertisers to sign up easily online themselves. This cut costs, brought midsize businesses into the fold, and gave Google an edge over similar services that had a lag between the time advertisers submitted copy and when the ads appeared. Text ads could be up and running on Google within minutes after a company provided its credit card number.

"To their eternal credit, Larry and Sergey both lighted on what was happening with [Overture's] business model and came to understand pretty rapidly what an attractive business that was," said Michael Moritz. "And it didn't take long for us to all understand that it made sense to go where the money was being spent. And it was being spent in the advertising business much more readily than in the licensing business."

Danny Sullivan saluted Google's approach on his Web site, an important vote of confidence that Sergey and Larry took very seriously. "The difference was that they understood you had to separate the editorial results from the paid results," Sullivan said. "And they always separated them."

As the months passed, Brin and Page had another breakthrough idea: ranking ads based on relevance, as they did their free search results. Instead of merely displaying an ad from the vendor willing to pay the most, Google ranked its ads based on a formula that took into account both how much someone offered to pay and how frequently computer users clicked on the ad. More-popular ads rose to the top; less-popular ones drifted downward. Brin and Page reasoned that the ads clicked on the most frequently were the most relevant. In other words, they trusted their users to rank the ads. It was consumer pull, rather than business push, that would determine where ads appeared.

"Making sure you showed the most relevant ads was a nice public relations kind of thing," Sullivan said. "You can go out and say, 'Our ads are more relevant because people are clicking on them.'" Sullivan also pointed out that Google would profit more by putting the most popular ads higher on the list, where they would be seen and clicked on by even more people. "This approach to advertising was home-grown at Google," he said.

Google had not been first in search, but they had come up with a way to produce superior results. They had not been first in search advertising, but they had done it cleverly and, they argued, with more integrity than Overture. And Sullivan was on their side, noting that the decision to rank ads in this way was better for all sides.

Still, Brin and Page felt that their search engine was at risk. Google had a growing brand name and millions of computer users turned to it for search daily, but it had no customer "lock-in" through site registration or an email service. What if someone else came up with better search technology? They had to remain focused on improving search through innovation. Google's results were better than those generated elsewhere, but they did not always answer the question on the user's mind. "The great thing about search is that we are not going to solve it any time soon," Page said in early 2000. "There are so many problems and failings. I see no end to what we need to do. If we aren't a lot better next year, we will already be forgotten."

Google's superior search delivered a big enough audience for businesses to target their advertising, but how well it would all work remained to be seen. By the middle of 2000, Google was handling 15 million searches per day compared to the 10,000 of a year and a half earlier. Computer users were flocking to Google, but would they start clicking on ads too, enabling businesses to sell products and Google to sustain its financial model? Brin and Page remained confident, but skepticism abounded about the company's ability to make it as a business that gave search results away for free and refused to accept both banner ads and paid placement in search results.

In December of 2000, *BusinessWeek* ran a story under the headline, "Will Google's Purity Pay Off?" In it, people wondered aloud whether Google could survive by focusing on search and users more than on money. "There isn't really good evidence, frankly, that companies focused purely on search, as Google has been, can support themselves with that model," said Marc Krellenstein, chief technology officer of search engine competitor Northern Light.

However, Google was now winning awards for the quality of its search results, and the media was taking note. Brin and Page had found a way to accommodate advertising without subtracting from the user's experience, and they held fast to their principles regarding the mixing of free results with paid ads. "When somebody searches for 'cancer,' should you put up the site that paid you, or the site that has better information?" Brin asked rhetorically. As editor of Search Engine Watch, Danny Sullivan was increasingly sought-after to provide expert commentary on the unfolding approaches taken by various search engines. In Google's case, Sullivan had chronicled it all on his Web site in bits and bytes, from Google's beginning, to its growth, to the changes in the dynamics of advertising. He had witnessed Google's rise in another way, too, as the company's presence at Sullivan's search conferences grew in tandem with the shows themselves. From a small booth at the original 1999 show, Google had become a main attraction on the exhibit floor, just as Danny Sullivan, the soft-spoken search impresario, had earned himself a spot at the center of a new industry.

Even as he enhanced his credibility as an impartial observer, Sullivan maintained very close ties to Google and the brain trust that its founders were assembling. "Google is very special to me," Sullivan said. "Most of the players were out there and existed before I got into it. With Google, it was brand-new—and I was a person coming in and watching them from the very beginning."

CHAPTER 8

A Trickle

Sergey Brin and Larry Page were perfectly positioned to pounce when the Internet stock bubble burst in 2000, prompting firings and bankruptcies all around Silicon Valley but not at Google. For a healthy, growing start-up, the timing could not have been better. The pair found themselves atop one of the only firms hiring amid the market crash that destroyed legions of public technology companies. As a private enterprise immune to the poison from Wall Street, Google now had access to outstanding software engineers and mathematicians who suddenly found themselves unemployed or holding a pile of worthless stock options. The pool of talent presented a one-time opportunity to add enormous brainpower and depth to Google that would have been impossible under normal business conditions. While many of its biggest competitors in the search engine field were in dire financial straits, with no hope of recovery, Google moved to a larger headquarters in Mountain View.

Inside, the playful atmosphere contrasted sharply with the shattered world that surrounded it. It was so distinctive that Stanford's Office of Technology Licensing, which had licensed the search patent to Google, described the fun-filled environment at the firm by noting its pool-cue-carrying programmers, its jelly-bean-eating assistants, and its satisfied computer users, who loved the simplicity, speed, and relevance of Google's search results. Brin and Page had spent judiciously when building the computer infrastructure that powered its business, but they spared no expense when it came to creating the right culture inside the Googleplex and cultivating strong loyalty and job satisfaction among Googlers. The artifacts of that culture—brightly colored medicine balls, lava lamps, and assorted gadgets and toys here and there—gave the business the appeal of a vibrant college campus. All of this, they believed, would pay off handsomely in the long run.

The 85 employees who now worked for Google CEO Larry Page and President Sergey Brin labored long hours but were treated like family. They were fed like family as well, with free meals, healthy juices, and snacks in abundance. Googlers also enjoyed a bevy of conveniences like on-site laundry, hair styling, dental and medical care, a car wash—and, later, day care, fitness facilities with personal trainers, and a professional masseuse—which virtually eliminated the need to leave the office. Beach volleyball, foosball, roller hockey, scooter races, palm trees, bean bag chairs, even dogs—it was all part of making work fun and fostering a creative, playful environment where Google's employees, most of them young and single, would want to spend their waking hours. Google would even go on to charter buses with wireless Internet access so that Googlers who commuted the hour from San Francisco could be productive, putting their energy into their laptops instead of worrying about how they would get to work.

Despite these unconventional approaches, however, the elements of a bona fide business strategy were taking shape.

With its search engine, Google had a best-of-breed product, a source of revenue through advertising, and a potent brand that conveyed not just excellence but a sense of fun and integrity too. The company also had the people and technology to scale up rapidly on all fronts, from computer infrastructure to ad sales and support staff to new product offerings. At the same time, its founders refused to veer from their total focus on users and the quality of search results, ignoring pressure from their investors to grab whatever dollars they could in a declining overall market. "Breaking a basic rule of advertising, its main page—the one most viewed by users—is devoid of ads," noted Stanford's licensing office in literature about its client. The company's most valuable online real estate remained unspoiled by ads of any sort, so it could load more quickly. Search results pages also loaded swiftly, since Google stuck to showing only targeted text ads, not bandwidth-hogging multimedia spots.

Not surprisingly, Google also eschewed flashy, untargeted ads when doing its own promotions. Instead, Brin and Page continued the word-of-mouth campaign they had begun while students, having loyal users email search results to friends, doing inexpensive promotions like giving out Google pom-poms at Stanford's home football games, and opening an online shop selling caps, T-shirts, lava lamps, and other items with the Google logo. They believed enough in the superiority of their search service to feel that average users would be eager advocates on their behalf, happy to "tell a friend" about the new tool they had discovered. The leverage of the powerful computing system they assembled was epitomized by those early days when they had only five employees and roughly one million people were using Google. Back then they had to be sure that searchers couldn't get the company's phone number, since they had no way to handle call volume. They proved to be masterful at branding, not through any grand ad strategy, but simply by letting their users and the news media carry their standard. "Their service was so good they grew organically," said Peter Sealey, a former marketing executive at Coca-Cola.

While Google was spreading its wings, a major potential competitor was getting its clipped. In June 2000, Microsoft, the company that arguably could have put together the most aggressively competitive search engine to Google's, lost a major federal court case in Washington DC when Judge Thomas Penfield Jackson found that its bundling of the Internet Explorer browser within the

Windows operating system violated antitrust laws. During the trial, Microsoft founder Bill Gates was depicted by prosecutors as a bully and a monopolist. This was the Gates and Microsoft that many in the software industry had come to know and despise, and they enjoyed seeing the billion-dollar behemoth get its comeuppance.

Once again, Google was a beneficiary of events and timing. Engineers who once longed to work for Microsoft came to see it as the Darth Vader of software, the dark force, the one who didn't play fairly. By contrast, Google presented itself as a fresh new enterprise with a halo, the motto Don't Be Evil, and a pair of youthful founders with reputations as nice guys. Its clean interface made people feel good about the product, one that always seemed to work properly—a far cry from the frequent reboots and ominous "Fatal Error" messages familiar to Windows users. In addition, Google had an idealistic mission: to make all of the world's information freely accessible and useful. The sharp contrast aided the search engine's recruiting, even as Microsoft, stung by bad publicity, pulled back from entering some new markets. While Microsoft was appealing the federal court ruling, it certainly didn't need or want the Justice Department, the European Union, or any other courts slapping it down again by accusing it of competing too aggressively.

Untarnished and increasingly beloved, Google gained further momentum when a major study named it the leading Internet search engine, with 99 percent of users identifying it as superior to the competition. It also moved to lock up the all-important university market, giving colleges its familiar colorful logo and search box to put on their Web sites and, in the process, breeding new users among well-educated students, faculty, and alumni. In a May 2000 profile, *The New Yorker* magazine described Google as the "search engine of the digital in-crowd." And that same month, *Time Digital* gave it the ultimate compliment: "Google is to its competitors as a laser is to a blunt stick."

As other tech companies in the Valley were shutting down, Brin and Page took their show on the road. "Hold on to your beret.

Google will soon be going French. And German. And Italian, Swedish, Finnish, Spanish, Portuguese, Dutch, Norwegian, and Danish," they wrote in an email to friends. "With so many of our loyal users around the world, it only seems fair to offer our search services in a variety of linguistic flavors." The company was translating the Google site into other languages and starting up Web sites in other countries to handle its growing international traffic with a more personal touch. It also began introducing wireless search capability, so cell phone users could Google on the go.

Then the company embarked on a mission to push its business and brand out to Internet users, rather than waiting for users to come to the Google.com homepage. In its new program, publishers of news, shopping, and other Web sites could sign up to add a Google search box to their own sites, giving their users access to Google search while earning money for the referrals. It was a change in trajectory that had plenty of upside for both Google and Web publishers of all sizes. Suddenly, Web sites—first in the U.S. and soon thereafter anywhere in the world—could enhance their offerings with high-quality search and be paid for it rather than having to pony up. From Google's perspective, it was akin to a television network like NBC or FOX spreading its brand name by delivering programming to affiliates across the country. The strategy had worked for the major TV networks for decades, and this new approach was to give Google unprecedented contact with millions of Internet users and Web site owners around the world. It would also boost exposure of the brand name at the very moment that its competitors were fading from view. To ensure maximum exposure and reach, Google had the wisdom not to be too greedy. "By signing up for our affiliate program, you'll be able to place a Google search box on your site and begin receiving 3 cents for each search you send our way," Brin and Page announced. "It's our way of saying thanks to all of you who have been spreading the word about searching the Google way."

On June 26, 2000, Google took a giant leap toward universal recognition by inking a pact with Yahoo to provide the Internet powerhouse with Google-generated search results. The deal vastly expanded its presence and profile on the Web, exposing it to millions of additional users daily. As one of the oldest, best known, and busiest sites on the Internet, Yahoo was an important customer, and its decision to make Google its new behind-the-scenes search provider, in place of incumbent and rival Inktomi, had important ramifications for the young company's future prospects. Yahoo officials said they chose Google for an array of reasons related to its search expertise and its focus on giving end users more comprehensive results.

"Yahoo selected Google because they share our strong consumer focus," said Yahoo president Jeff Mallett. "Google has clearly demonstrated its ability to scale with the rapid growth of the Web, making it a particularly good match for Yahoo as we continue to expand our global presence."

Sergey Brin called the deal "a significant milestone for Google and a strong validation of our business strategy." The pairing seemed natural given that the two companies' founders knew each other, had all been Stanford Ph.D. students, and had received backing from the same venture capitalist, Michael Moritz. The deal had special significance for Larry Page, since his brother, Carl Jr., also was in serious negotiations with Yahoo over a major business transaction. The following day, June 27, Yahoo announced plans to buy eGroups, a technology firm that Carl Page had cofounded, for $413 million. It had been a great week in the relationship between Yahoo and the Page brothers, and a superb week in the relationship between Google and Yahoo. To top it off, Google announced it was now the world's largest search engine, with more than one billion pages in its index of Web sites. This meant that it not only was the fastest and most relevant search engine but also the most comprehensive. "Now you can search the equivalent of a stack of paper more than 70 miles high in less than half a second," Page said. "We think that's pretty cool."

By early 2001, Google was performing a staggering 100 million searches per day—10,000 every second. It was also entering the

American lexicon as a verb, a trend documented by a *New York Observer* article that chronicled New Yorkers googling each other before dates. Though less frequently discussed, ego or vanity searches—people googling themselves—were also on the rise. It was human nature. To many people, a validation of their importance rested upon showing up in a Google search.

For Google, validation of its superiority rested with search guru Danny Sullivan. "The acknowledged expert on all things search is Danny Sullivan, editor of Search Engine Watch," Brin and Page wrote to friends in January. "After polling the readers of his Search Engine Report, Danny named Google 'Outstanding Search Service' and 'Most Webmaster Friendly.' This honor is especially gratifying given Mr. Sullivan's intimate knowledge of the intricacies of every kind of search service available online."

On the business side, Brin and Page, with help from their head of sales, Omid Kordestani, attracted America's biggest retailer, Wal-Mart, and a major automobile manufacturer, Acura, to the new medium. The big firms joined thousands of medium-size businesses that advertised their wares on Google. Rather than seeing the ads as something evil, Brin and Page began to view the relevant ads generated by Google as an important part of the information provided to computer users conducting online searches.

"What's the secret behind the rapid growth of Google's advertising program? The answer is Google's unique approach," they said. "Google runs only keyword-targeted text ads. That means you don't see the ad unless you're searching for information on that specific topic. And because there are no animated banners competing for attention, the text ads are read carefully by users, who frequently find them to be as valuable as the actual search results."

How would Google capitalize on this trend to become a megamoney maker? Enter Yossi Vardi, the Israeli entrepreneur and venture capitalist, and Eric Schmidt, the computer scientist who would become Google's chief executive officer. This pair of experienced business executives independently generated ideas that would dramatically increase the company's sales and profits—and the Google Guys followed their advice.

Rather than running only a few text ads atop the free search results, as they had been doing, Vardi suggested that they divide the Google search results page with a vertical line. Devote two-thirds of the page to free search results, Vardi told them, and designate one-third of the page, to the right of the results, for text-based ads. The suggestion, which Brin and Page accepted after bouncing the idea off numerous people, dramatically increased the available space that Google set aside for ads. It also increased the prominence of ads for computer users to click on. They had initially worried that the move would compromise the perceived quality of search results, but Vardi, an experienced veteran of business, finance, and technology, persuaded them that as long as the difference between the free search results and the ads was clearly delineated, the integrity of the results would be preserved, the page would still look clean, and revenue would pour in much faster.

Eric Schmidt would ask an important question, soon after his arrival as CEO, that Sergey and Larry had not explored: where were Google searches originating, and where were the company's ads coming from? The answer provided a road map for a new ad sales strategy. While 60 percent of its search requests came from outside the United States, the company earned only five percent of its revenue from ads that originated outside of North America. The Google Guys had put tremendous emphasis on global branding and making Google searches readily available in foreign languages, but they had done nothing specifically to generate ad sales abroad. Schmidt knew exactly what to do. He directed supersalesman Omid Kordestani to go to Europe and not return until he had hired key executives to run the company's European sales operations. Schmidt joked with Kordestani that he was about to rack up megamiles on United Airlines, flying between the West Coast and Europe to build a major overseas sales force for Google. Kordestani succeeded. Before long, foreign sales offices were up and running in London and Hamburg, as well as in Tokyo and Toronto.

Inside the Googleplex, innovation continued. One engineer devised a way for searchers to find a phone number on Google by simply entering someone's name and zip code into the search box. Another came up with a way to catch spelling errors. After a person typed words in the search box, if one of the words was spelled incorrectly, Google automatically asked, "*Did you mean xxx?*" This was as close as it had come, thus far, to reaching into the minds of searchers to divine not what keywords they actually typed, but what they *meant* to type.

Google also launched an extraordinary new feature that would revolutionize many aspects of Internet usage over time. Called Google Image Search, the service included millions of photographs and other graphics available with the click of a mouse. All a user had to do was type a name or description into the Google Image Search box and instantly photographs and more appeared. Image Search, with its many uses, was a global hit. It also showed that the Google search model had major expansion opportunities. Google's initial index of 250 million images was so vast when it was launched in the summer of 2001 that it had no real competition. "If a picture is worth 1,000 words, what about a million pictures? Or to be more precise, 250 million pictures?" Brin and Page wrote in announcing the feature to friends. They also warned the public that images of adult content might appear unexpectedly. "You should be aware that the results you see with this feature may contain adult content. Google considers a number of factors when determining whether an image is relevant to your search request. Because these methods are not entirely foolproof, it's possible some inappropriate pictures may be included among the images you see." And harkening back to the notion of googling someone before going out on a blind date, people now had the power not only to read about their potential suitors but also to see what they looked like.

When terrorists attacked the United States on September 11, 2001, Google's search traffic surged. "Many important news sites were overloaded by heavy traffic and could not serve an information-starved public," Brin and Page noted. "Google did its best to fill the void by putting up cached [stored] versions of news stories on the Google homepage, and Google continues to maintain an extensive set of links to major news sources around the world." On routine days, and on those of extraordinary events as well, Google was woven into the fabric of American culture—and increasingly, with its availability in 66 languages, the global community as well.

As the end-of-the-year numbers came in, it appeared Brin and Page's business strategy was paying off. The three-year-old company was in a much better position than many of its dot-com brethren. A relentless focus on innovation and cultivating the right corporate culture spawned a barrage of new products and areas for potential business growth. Traffic continued to grow exponentially. And ads had begun generating revenue, though it was still just a trickle.

Michael Moritz, a venture backer and Google board member, liked what he saw. He knew Google had the best search technology on the market and the right pieces in place. The rest would come in time. "Revenue always begins in dribbles," Moritz said. "It is like rain showers. You always get a few drops of rain before the downpour."

Larry Page joked that the reason Google made becoming profitable such a priority was that Sergey wanted to be able to impress women on dates, and that being president of a money-losing dot-com no longer had appeal. "In Palo Alto in 2000, a huge number of people were presidents of money-losing dot-coms, and so they would not call him back," Page said. "And he thought, 'If only I were president of a moneymaking dot-com, things would be very different.'" Brin and his wingman Page got their wish in 2001. With a stack of text ads now running to the right of Google's search results, giving users more options to click, the stream of revenue from searches that year was enough to earn Google its first annual profit. It amounted to $7 million.

CHAPTER 9
Hiring a Pilot

Eric Schmidt had no interest in visiting Google even as he walked in to meet with Sergey Brin and Larry Page in December 2000. The first thing he noticed was that they had projected his biography on the wall. He had heard that Google was a flaky place, and this seemed to confirm it. "I thought that was really odd," he said. He had done his best to avoid this meeting, but John Doerr of Kleiner Perkins, the most powerful venture capitalist Schmidt knew, wouldn't quit asking him to meet the Google Guys and at least talk about playing a role on the management team. Schmidt respected Doerr and valued his relationship with him immensely. If Doerr had not been involved with Google as an investor and board member, he would have blown off the meeting completely. Even so, he had done his best to put it off ever since Doerr approached him at a political fundraiser for a local congressman in October of 2000.

"Go talk to Google," Doerr said.

"Nobody really gives a shit about search," Schmidt replied.

"Go look at Google," Doerr said again. "This is a little jewel that needs help in scaling it."

In the small world of technology companies and upper echelon financiers, Schmidt didn't know anybody who had a better record of success over time than Doerr. That was a relationship worth preserving, even if meeting with Brin and Page turned out to be a total waste of time. Schmidt, the chief executive officer of software maker Novell, wasn't looking for a new job, though he knew he would need one soon after Novell completed a merger. But even if he had been ready, Google was not the place this Ph.D. business executive would have turned. Doerr's enthusiasm notwithstanding, Google was nothing more than a search engine at a time when the buzz in Silicon Valley was that search engines were dead and the all-inclusive Web portal was the business model of choice. Schmidt believed in the gospel that a handful of portals offering news, weather, shopping, and email would be the place people lingered on the Web.

He also had a hard time understanding why Doerr was so enthusiastic about Google. Maybe, just maybe, he was worried about getting his investment back, and hoped Schmidt could ride to the rescue and do some damage control. In any event, Eric Schmidt reluctantly walked into the room, ready at last to meet the young entrepreneurs, fulfill an obligation to Doerr, and then get back to work at Novell.

Sergey and Larry had as little interest in meeting Schmidt as he had in meeting them. He was no more than the latest in a series of technology chieftains that they would be obliged to waste their time with in order to appease Doerr and Moritz, their venture capital backers. Brin and Page planned to send Schmidt on his way back to wherever he had come from, just as they had everyone else. The truth was, they still didn't want anybody looking over their shoulders at Google—and the last thing they needed or wanted was a bean counter. Putting corporate-type controls in place could only hurt, stifle innovation, hamper progress, and lead to alarming reports back to Kleiner Perkins and Sequoia Capital about how they were wasting money. Some senior executive from the world of business and technology would never be able to understand the culture of the enterprise they had created, since it was more like a graduate school program on a university campus than a business on its way to an initial public offering, the coin of the realm for venture capitalists looking to cash in.

Sergey and Larry took pride in their independence. The two venture firms that had invested $25 million in Google had gotten nothing in return thus far except frustration, headaches, and sporadic updates from the pair, who had done everything possible to maintain absolute control. But the firms were insistent that they hire someone more experienced and older than them, someone who knew how to run a company and could be its public face when it came time to cash in on Wall Street.

The guys were fully aware of what was going on. They had played one firm against the other as masterfully as a skilled poker player bidding and bluffing his way to capturing the chips, regardless of whether he was holding the best or worst hand at the table. Sergey, a student of human nature, and Larry, who had the benefit of learning from his older brother Carl's experiences, knew that the most painful experience a venture investor could have was passing on a potential investment that turned out to be a blockbuster. That was one of the things that kept wealthy men in the game.

Eric Schmidt didn't know it when he walked in to meet with Brin and Page, but John Doerr had been looking for the right person to bring in as CEO of the search engine for 16 frustrating months. While Brin and Page had made a commitment to hire someone they approved of as chief executive officer, Doerr was beginning to feel that nobody he identified would be good enough to please them. The person would need the right combination of personality and smarts to meet their high standards—and be willing to check his ego at the door. They had rejected one candidate after another. As Brin and Page saw it, they would remain in charge of Google, not some new CEO hired from the outside and forced on them by Doerr. From their perspective, everything was fine at Google. They had done their best to discourage many candidates Doerr sent along from wanting to come to work with the two of them.

The room that Schmidt walked into, where Sergey and Larry had food on a tray and his background projected on the wall, was in the old Sun Microsystems Building in Mountain View. Schmidt was a former chief technology officer of Sun, but he had long since moved on to tackle new challenges at Novell. Now, almost as soon as he sat down, Sergey laid into him about what he termed the "stupidity" of the strategy Schmidt was executing at Novell. "I argued back hard," Schmidt recalled. "We argued for at least 90 minutes." Back and forth they went, debating, disagreeing, and intellectually dueling. By the time he walked out, Schmidt, who previously had laid the groundwork to leave Novell over a period of months, had two thoughts about their discourse: it was the best argument he had had in a very long time, and he sensed he would end up involved with Google in some way.

For their part, Brin and Page liked Schmidt better than the other candidates they had met. They were under increasing pressure to bring someone in who could handle the internal management side of the company. At one point, Moritz had threatened to demand repayment of Sequoia Capital's $12.5 million investment if they did not fulfill their verbal pledge to hire a first-rate chief executive officer. It had been a heated exchange, but it didn't motivate the Google Guys, who dismissed the threat. To Moritz, it felt as if they were resisting his directives like a pair of adolescents challenging parental authority. "If Larry and Sergey were given instructions by a divine presence, they would still have questions," he said.

Recognizing the magnitude of the challenge, Doerr decided to take a different approach. He arranged a series of meetings for Brin and Page with heavyweights in the world of technology whom they respected, including Intel chairman Andy Grove. Doerr hoped that over time these meetings would persuade the independent-minded entrepreneurs that they would do better and be happier in the long run, as Google grew, if they focused on intriguing problems and left the mundane tasks of running a business day-to-day to somebody else. It seemed better than trying to persuade them to honor a verbal pledge they had never considered important in the first place. But it wasn't clear to Doerr how well his strategy was working. After the guys had a conversation with Jeff Bezos, according to an article in *GQ* magazine, the Amazon CEO told Doerr, "Some people just want to paddle across the

Atlantic Ocean in a rubber raft. That's fine for them. The question is whether you want to put up with it."

Still, Doerr was optimistic that Brin and Page would eventually come around, and he sensed that Schmidt had the right blend of personality and background. One thing the pair did like about Schmidt was that he not only had experience as a CEO but was also a computer scientist. Like the best people they hired at Google, Schmidt had academic credentials and a scientist's love of research. He held a Ph.D. in computer science from the University of California at Berkeley, and had earned an undergraduate degree in electrical engineering from Princeton. He had done research at the Xerox Palo Alto Research Center—known by the acronym PARC—as well as at Bell Labs. He wasn't afraid to speak his mind and wasn't intimidated at their first meeting. He had also been to Burning Man. In short, Schmidt had passed their litmus test.

Schmidt also had something else others might view as a weakness but Brin and Page saw as a strength: he had failed at something. While at Sun Microsystems, he had challenged Microsoft by leading the development of Java, an independent programming language, and he had defined the company's Internet strategy. Although the effort proved largely unsuccessful, it showed that Schmidt was not afraid to take on Bill Gates and Microsoft's computer-based operating system with software that gave businesses and people a choice rather than having a system forced upon them. This took a certain independence that Brin and Page admired, and it also meant that Schmidt recognized what Sun had done wrong in challenging Microsoft's dominance. For the Google Guys, the advantage there was that rather than making mistakes themselves as they built an Internet-based software alternative, they could learn from the strategic and tactical errors that Schmidt and Sun had made.

Their initial meeting ended inconclusively, but a courtship had begun.

Sergey and Larry liked Schmidt as well as they were going to like anyone in the role of Google CEO. But one thing was certain: they had no intention of lavishing him with cheap stock options that would someday be worth big money, without extracting a hefty commitment. They wanted proof that he, too, was emotionally and financially invested in the company that they had founded and built.

Sergey and Larry called him to talk things over.

"What would you like to do?" the Google Guys asked Schmidt.

"I am busy selling a company," Schmidt reminded them. He had no intention of walking out at a moment when he was trying to guide Novell through the sale process.

"I am happy to be chairman of Google," he then said, since this would not entail day-to-day responsibilities, "and be CEO at some point in the future."

"We don't need you now," Larry told him, "but we think we're going to need you in the future."

"I agree that as the company grows my experience would be helpful," Eric replied.

The parties hung up and took a deep breath. Behind the scenes, Doerr pushed both sides to find a way to make this work. He sensed it was the right match. "John's role was fundamental," Schmidt said. "When you looked at Google from the outside, it looked like Larry and Sergey having a lot of fun with smart people. Mike [Moritz] and John brought gravitas to the effort. That should not be trivialized."

Next came serious negotiations over Schmidt's compensation package. Larry and Sergey were insistent that he invest some of his own money in the company. He was equally insistent that if he was going to become chairman first, and then chief executive officer, he had to have the opportunity for plenty of upside in the form of stock options, the currency of choice in Silicon Valley. Schmidt knew the company was cash-poor and idea-rich, so he didn't push for a big annual pay package. He would make plenty of money from selling Novell. But by ignoring the prevailing wisdom about search to go work for Google, he was putting something far more valuable on the line: his reputation.

They went round and round over stock options, which Larry and Sergey had doled out generously to their friends from Stanford and elsewhere who had come to work with them back when the idea seemed even riskier. But if Schmidt was going to take the job as chairman and then CEO, he wanted options that could be converted into millions of shares of stock, making him the biggest employee shareholder.

Two things helped bring the deal in for a landing. First, Schmidt showed his commitment by agreeing to pay $1 million of his own money to buy preferred stock in Google. He did so at a time in early 2001 when the company was running short of cash, so the money truly served a business purpose. Second, Brin and Page recognized that John Doerr had some leverage over them. More than a year had passed since he and Moritz had invested the $25 million in Google, so they could say that the pair had breached their promise to hire a chief executive officer. The venture firms could now demand their initial investment back, which Google could not afford either financially or in terms of its reputation.

Schmidt reached an agreement with Brin and Page in January 2001, and the deal was signed two months later, in March, after all the financial and legal terms had been finalized. From March until July, Schmidt served simultaneously as chairman of Google's board and CEO of Novell, spending some time at Google on top of his day job at Novell. "You can't be CEO and not show up," Schmidt pointed out. In the second week of July, when the Novell merger was completed, he was appointed chief executive officer of Google Inc.

"Not only was I CEO and chairman, but I was also an investor of real money," Schmidt recalled. "The company literally needed the cash. And they wanted to see a real commitment."

Schmidt arrived at Google to find a technology firm that nearly three years after its founding was being run by technologists who put enormous time into people and products and users, but spent as little money and time as possible on the details of internal management. He moved into an eight-by-twelve-foot office near the pair, who shared a larger office full of people and toys and computer equipment and other random items. "The place was always a zoo," Schmidt said. He knew what he needed to do, but he also had to persuade Sergey and Larry to accept the necessity of building a business infrastructure. For instance, the financial record-keeping and payroll systems were being run using off-the-shelf software from Quicken, the kind people use to do their own income taxes or operate a very small enterprise. "That was fine for a start-up, but not for this company with 200 employees and $20 million in revenue," said Schmidt.

This turned into a defining battle. Schmidt wanted to bring in a major business and financial record-keeping system from Oracle; that was his job. But Larry and Sergey thought it was a bad idea, a complete waste of money. Eric found himself challenged on doing the basic task he had been brought in to accomplish. "That was a huge fight," Schmidt recalled. "They couldn't imagine why it made sense to pay all that money to Oracle when Quicken was available."

Despite conflicts like this, Schmidt had the right sensibility and touch with Larry and Sergey. He knew when to push, he knew when to back off, and he knew when to make light of their differences. The more time he spent at Google, the more impressed he became with the culture they had created, and the clarity and sense of shared mission that pervaded the company. They had a broad vision. His job was to put that vision into a framework that would give it the best chance to produce tangible financial results.

"The underlying structure and strategy and culture were good," Schmidt said. "The most apt description of what I did in the first year or two was put a business and management structure around the vision and gem that Larry and Sergey had created."

This sounded simple enough, but it wasn't easy to achieve, since Schmidt didn't want to do anything that would alter the company's DNA or destroy his relationship with Larry and Sergey. For Schmidt, an engineer, a business executive, and an avid pilot, it often felt like trying to rebuild an airplane in mid-flight without touching down long enough to refuel. Eventually, however, the trio, with the help of an outsider, Intuit CEO Bill Campbell, whom Doerr brought in to guide them, came to function more as a team. Schmidt learned which battles to fight, when to look the other way, and how to forge trust and implement some natural division of labor so that decisions got made. As it turned out, Sergey was a gifted deal-maker, Larry was the deepest technologist of the three, and Eric focused on the details of running a business.

Still, Brin and Page liked to kid around. The guys unleashed a series of practical jokes to test Schmidt's mettle, disrupt his momentum, undermine his authority, and make sure he understood his place in the pecking order. "When I got here," Schmidt said, "the company had credit cards that didn't get billed to individuals. Larry and Sergey just gave credit cards out. The first thing I did was cancel all the credit cards except one, which Larry and Sergey controlled. They gave their card to other people to use to buy stuff, just to spite me. One day a telephone booth showed up in my office. I said, 'Who bought the telephone booth?' We tracked it down to somebody who had their credit card number. It was very entertaining. One day these massage chairs showed up. Who bought 'em? I don't know. A little bit of mischief goes a long way."

As Google grew, it became clear that John Doerr's role in encouraging Page and Brin to hire Schmidt mattered more and more. So did his insistence that Bill Campbell spend time advising the trio as an outside consultant and coach; he helped them function well together. The Google Guys were receiving much more than money from Doerr and his firm. They were getting the guidance they needed to operate a growing, privately held business in a professional manner, while retaining the sense of innovation and entrepreneurship they held dear. And at last they were accepting the unorthodox hierarchy of having a trio at the top.

In that trio, even though he held the ranking title of CEO, Schmidt would soon find out that he could be outvoted.

CHAPTER 10
You've Got Google

Google attained new financial heights in 2002, making it clear for the first time that its business potential could rival its revolutionary influence as a search engine. America Online, which owned Web properties that connected more than 34 million subscribers to the Internet, adopted Google as its search engine of choice on May 1. From that point forward, AOL users had a small search box on every page that said, "Search Powered by Google." AOL's size extended Google's reach more than any other partnership the company could have entered into during that period. It came about in part because Sergey Brin and Larry Page had focused on putting users first and providing them with the fastest and most reliable results, and it was also a competitive coup, since Google triumphed over rival search advertising company Overture, which had been providing ads to AOL.

The initial idea to add Google search to the AOL service originated with America Online cofounder Steve Case. A devoted AOL user, Case found himself turning to Google to do searches for information of all kinds, and realized he was journeying there because it provided him fast answers to his questions. If he made the journey, other subscribers would too. Though he was not involved in day-to-day management of AOL at that time, he suggested to

AOL executives that they strike a deal to make Google the official search engine of America Online's flagship service.

Winning the AOL business against both Inktomi, which provided search results, and Overture, which provided the search-related ads, was not as easy as gaining Case's loyalty as a Google searcher. The only way to get AOL to switch, it appeared during hard-fought negotiations, was to provide it with a very large financial guarantee, running to many millions of dollars. In exchange, Google would take over both search and search-related advertising on America Online and share revenue. AOL also demanded stock options as part of the deal.

Brin and Page, eager to grow Google, were prepared to pay virtually any price to win AOL's business, but their new CEO, Eric Schmidt, was wary, concerned that the kind of multimillion-dollar guarantee America Online was demanding could put Google out of business if it ran out of cash. It wasn't long ago, Schmidt recalled, that Google had $9 million in cash and $9 million in debt on its books. Conditions could change overnight, and over the years, he had learned the importance of a cash cushion. A ferocious debate ensued at the top of Google, with Schmidt opposed to AOL's demands and the founders joined at the hip in favor of striking a deal. By itself, the proposed partnership with AOL was so large that Schmidt discussed the possibility of doing a special private round of financing to ensure that the search engine did not run out of cash as a result of the deal, and then crash and burn.

"I was terrified," Schmidt said. "Larry and Sergey and I argued hard. I understood that if you ran out of cash, you were done. They were more willing to take risk than me. They turned out to be right."

Google's alliance with AOL once again pitted it against Microsoft. For years, Microsoft had threatened to wipe out the AOL service by spending heavily to promote its competing Internet service, MSN, and by offering free email through its Hotmail subsidiary. Now AOL, which had purchased Netscape only to see it wiped out when Microsoft offered its Internet Explorer browser free to computer users, filed a major lawsuit against Microsoft, seeking damages for unfair, anticompetitive practices. Microsoft, for its part, selected Overture to provide ads for its fledgling search service. In the battle between Google and Microsoft, sometimes overt and sometimes covert, both parties were doing their utmost to gain an upper hand, their rivalry simmering. Meanwhile, one of Eric Schmidt's former employers, Sun Microsystems, filed a civil antitrust lawsuit of its own against Microsoft. Most important, the attacks on Microsoft hurt the giant company's image just as Google was burnishing its own through a series of new partnerships.

"By far the biggest buzz at the Googleplex these past weeks was the announcement of Google's strategic partnership with America Online," Brin and Page wrote in an email to friends. "According to Bob Pittman, COO-elect of AOL Time Warner, 'Google is the reigning champ of online search. We're committed to providing AOL members and Web users with the very best in online tools, content and convenience, and we're very pleased to bring Google's popular platform to our users.'" Overture stock lost nearly one-third of its value on the news that it had lost the AOL business to Google. "This is an important endorsement of our advertising model and the way our customers view us," Schmidt said. "Our business outlook is very good and this AOL deal will help us."

Search expert Danny Sullivan, looking at the way Google knocked both Inktomi and Overture out of the box at AOL, marveled at its growing presence. "It's an incredibly huge victory for Google," Sullivan said. "It's an easy all-in-one solution."

In addition to the deal with AOL, Google entered into a pact to provide search on EarthLink, another Internet provider. Google also closed a three-year deal valued at $100 million with Ask Jeeves, a rival search engine, to provide it with text-based advertising. Jeeves would continue to produce its own search results based on proprietary technology, but by partnering with a competitor, Google showed signs of growing up.

That spring, it adopted a new policy of charging advertisers only when their ads were clicked. It was designed to give them greater control over how much they spent on ads. Overture, stung by having lost important accounts to Google twice, filed a patent infringement lawsuit charging that Google's search-related ad business improperly used registered technology developed by Overture, a pioneer of the click-based system. Google said the lawsuit had no merit, but Overture vowed to see it through. Meanwhile, Google's continuing role in real life and on television gained further currency when one of the characters on the popular NBC television show *The West Wing* suggested that a fellow worker try to find information by doing a Google search. Unlike some other companies, Google paid nothing for the mention on the hot program.

"Ultimately, the Google story is all about product," said Steve Case, cofounder of America Online. "It was better. It built some buzz by saying it was better. People used it because it was better, so that made it better still. Because it was exciting and new and different, Google was able to attract a great team of engineers. It was a better way to search the Internet and extract information. That is the driver at the core."

Far from the view of ordinary Google users, the search engine operated a sophisticated 24-hour marketplace where thousands of words and phrases that people search for every day were bought and sold like goods and services. The name of an everyday item such as "pet food" might fetch 30 cents, while those with high profit margins such as "investment advice" could go for as much as three dollars. The prices were what advertisers were willing to pay for their ad to show up when a user searched for that term on Google. And how the prices were set in this electronic marketplace answered one of the biggest mysteries about how Google converted the clicks on its Web site into cash on its balance sheet.

Google made money every time a computer user clicked on one of the ads it displayed. But rather than being set in advance, the cost of running an ad on Google, or on its network of affiliate sites, was determined in a nonstop online auction. This way, Google ensured that it always received a competitive price for each of the millions of ads it displayed daily. The result was more money to Google's bottom line.

Businesses around the world employed people to spend their days in front of computer screens bidding in this electronic auction. Some used automated software to do the job; others hired outside marketing professionals who specialized in this type of auction. However it was done, the bidding took place behind the scenes, so all that Google users saw were the winning ads, stacked along the right side of their search results.

Antonella Pisani, an executive at computer and electronics maker Gateway, Inc., bid in Google's keyword auction daily. She thought of herself as a portfolio manager, dealing in search keywords and phrases rather than stocks and bonds, but nonetheless hoping to maximize her company's investment in online advertising. She juggled bids, for example, on the term "digital camera" and its plural, "digital cameras"—with the cost of a click on the second averaging $1.08 compared to about 75 cents for the first. The reason for the disparity, she said, was that customers who typed in the plural were more likely to end up as buyers. These sorts of nuances were learned with experience and they separated the most successful bidders from the rest of the crowd, but the basics of the auction were kept as straightforward as possible to entice the greatest number of bidders.

Google did not invent the concept of auctions for search terms. That distinction belonged to Overture—a company purchased by Yahoo, and Google's auction advertising rival. But as patent litigation by Overture over the systems' similarity moved through the courts, Google's virtual auction floor grew more and more crowded, and consequently more profitable.

Companies of all sizes were participating in these keyword auctions, spending anywhere from hundreds of dollars to many millions each quarter on Google. Turning traditional advertising on its head, these ad buyers determined the price they were willing to pay to get across their message, rather than having Google set the price, as TV networks and newspapers had long done. The self-service nature of the system and the low minimum ad prices enabled even small firms with no sales staff to jump in and out of the game. On Google, mom-and-pop enterprises had the same opportunity to reach millions of users as a Fortune 500 company.

An old adage was that companies typically spent twice as much as necessary on advertising but had no way to figure out which half to cut. Google saw this as yet another problem it could remedy through clever technology. Click-based pricing of ads made it easier for firms to measure the effectiveness of their campaigns, since they could track whether users who clicked on particular ads turned into buyers. (Clicking on an ad typically took a computer user to an Internet page where a purchase could be made.) If the ads were converting to sales, a company could increase the amount it was willing to bid or expand its campaign; if the ads were ineffective, they could be adjusted downward or pulled entirely.

In Google's nonstop auction, the minimum bid for a search term was five cents, but that was about the only price that was stable. The rest rose or fell depending on what companies were bidding at any given moment as they tried to tweak their positions on the results pages. One of the most expensive search terms was "mesothelioma," a type of cancer caused by exposure to asbestos. Top bids for the word hovered above $30 per click and came from lawyers aggressively vying for the chance to land a lucrative client.

However, being the highest bidder on Google did not guarantee that an ad would appear in a top-tier position. This is one of a number of differences in style between the rival search engines. Google ranks ads based on two factors: the price a company is willing to pay and how frequently computer users click on the ad. Thus, even if a company outbids others on a particular keyword, if consumers are not clicking on the company's ad, it will move down to a less prominent spot. Yahoo, by contrast, guarantees that the highest bidders for a word will show up at the top of the list of sponsored ads. Yahoo is "strictly capitalistic—pay more and you are number one," said Dana Todd, an executive with an interactive ad agency. "Google has more socialist tendencies. They like to give their users a vote."

Advertising on Google proved to be an extremely efficient way for firms to reach potential customers. Google offered narrowcasting, not broadcasting—it tried to reach the consumer at the point of decision about buying a product, rather than plastering ads in places with the right customer demographics. The online ad business grew rapidly as firms of all sizes around the world discovered how well the search-triggered ads led to sales. "It is a very efficient marketing program. You are capturing people while they are interested," said Gateway's Pisani. And as people's lives moved online more and more, advertising followed them there. In many ways, Google's financial success was the result of being a leader in figuring out how to make advertising on the Internet work, both for businesses and for computer users.

Over time, Google would loosen restrictions on the ads it allowed on its partner Web sites, to include those with images and photos, not solely text. And it would give major advertisers greater control over where their ads did and did not appear. But it would stick with its principle of combining an ad's popularity with how much a business was willing to pay. On the Google.com site, text-only ads in small square boxes would still rule.

For all the money and attention spent in the electronic auction, Google's free search results remained the best spot to be prominently displayed. Studies that tracked users' eyeballs showed that they first looked to the left side of the page, where the free search results were posted. An entire industry of search "optimizers" developed solely to tweak firms' Web sites so they would appear prominently in Google results. Consistently appearing high up in the free results, however, was trickier than winning a top spot in the ad auction.

Whatever the method, whether it was free results or ads, what counted the most for businesses was being near the top. As Frederick Marckini, chief executive officer of online marketer iProspect, said, "All research confirms that if you are not found on the first three pages of the search results, the top 30 matches, you have built a billboard in the woods. No one will find it."

With Yahoo, AOL, EarthLink, and Ask Jeeves as its partners, Google now had relationships with most of the biggest and best-known Internet properties. Its network of affiliates across the Web that featured Google search on their sites also proved to be lucrative ways to drive traffic and raise brand awareness. Over time, the number of sites with a Google search box would grow to a staggering 25,000, forming a money-generating network that was exceedingly difficult for anyone to replicate.

The Internet revolution, it turned out, was fundamentally changing the nature of advertising and the marketing of goods and services, and Google was poised to make the most of it. It had already helped to vault medium-size businesses, and consumers, into new positions of power before many Fortune 500 companies even realized that a transformation was taking place. Empowered as never before, businesses now had a way to target customers online in a direct and relevant way, and consumers could hunt for information and do research about products for free on the Internet, then click on ads if and when they were interested. And they could do it all conveniently, without the need to go to stores or wait in lines. The power was at their fingertips. It was a modern-day version of letting your fingers do the walking.

Google and its venture capital investors saw that the search engine was going to be a runaway financial success. The search engine generated $440 million in sales and $100 million in profits in 2002, although the world didn't know it, since Google was a private company. Virtually all its profits came from people clicking on the text ads that graced the right side of search results pages at Google.com and the pages of its many partners and affiliates. Because its financial prowess was hidden from the outside world, Brin, Page, and Schmidt were free to grow the company as aggressively as they chose, without having to look over their shoulders.

They stayed absolutely silent about their financial numbers to prevent others, especially Microsoft and Yahoo, from finding out how profitable their online search and advertising business had become. It would have prompted the others to invest big money in building or buying search engines of their own.

With this major head start, Google's hypergrowth would continue as companies shifted billions of dollars of their ad spending from television, radio, newspapers, and magazines onto the Internet. The onetime pure-play search company now sat at the epicenter of the entire Internet advertising economy.

CHAPTER 11

The Google Economy

Steve Berkowitz had an inkling about the raw, transformational power of the Google Economy and the Internet before he was recruited in 2001 to revive the flagging search engine Ask Jeeves. With the kind of business savvy that enabled him to see the next two or three moves on the chessboard, he relished the role of the underdog and wasn't fazed by the challenge of turning a drifting enterprise into a growing, profitable business. He had the confidence that comes with having done it before. From a standing start, Berkowitz had built the "For Dummies" book series from an obscure imprint of computer titles into a major force in the publishing industry. A technology outsider, Berkowitz intuitively grasped what many Web insiders completely missed: that the Internet had much in common with traditional media and could be viewed simply as another way of delivering content and advertising to users worldwide. His ability to peer into the future, and his understanding of the close relationship between branding and customer loyalty, translated readily from the world of book publishing to the Google Economy.

In his new post atop Ask Jeeves, Berkowitz inherited a well-known but financially ailing brand that consistently had over-promised, underperformed, and unwittingly undercut the immense value of its franchise and best-known symbol: a fastidiously dressed

British butler named Jeeves. The presence of the butler on the Ask Jeeves site gave it a human touch that appealed to something in the American psyche. It converted the typically impatient computer user into a more empathetic being who wanted the mythical British servant to deliver topnotch service and was willing to give him a second chance.

For all of Jeeves's genteel appeal and Berkowitz's business savvy, the reality is that both would likely have failed had it not been for the Google Economy. The success of the globally loved search engine was having a ripple effect on businesses that benefited from quality search, relevant advertising, and renewed faith in high-tech innovation. Google's record of creating thousands of jobs and pumping billions of dollars into the economy translated into innumerable opportunities for entrepreneurs and business moguls alike. This could only be seen clearly from inside Google's corporate headquarters, where globes with twinkling lights and backlit maps tracked the millions of searches that streamed in day and night from around the world, revealing the innate power and reach of the Google brand beyond anything the Internet or the business world had ever seen before. Traditional metrics used to gauge the popularity and profitability of a business missed the dynamic, self-reinforcing nature of Google's global network of users and advertisers, in part because the unconventional business model didn't lend itself to standard static analysis. Google was more like a snowball rolling down a black diamond ski slope, gaining in momentum, size, and velocity along the way.

The most popular new application on the Internet since email, Google helped spark new ideas, connecting and reconnecting business partners and friends, and empowering entrepreneurs to conduct market research, recruiting, advertising, and self-promotion that was inexpensive and instant. Without declaring such a purpose, it had become a business owner's best toolkit. In ways large and small, it was influencing a broad swath of the American and global business world. And not just business— universities, government, and research institutions felt the effects too. Money and ideas were changing hands at a faster rate than ever before, thanks to the information Google indexed and made available at lightning-fast speed. It was not an overstatement to call the young company, as a search engine, an advertising engine, and a growth engine, the catalyst for an important new economy.

Steve Berkowitz didn't fully understand the world he had stepped into when he arrived at Ask Jeeves in May of 2001 with a mandate to save the once high-flying business from the death spiral that had destroyed so many other promising dot-coms. But he had learned about Ask Jeeves at the time he cut a deal with the company that turned his Dummies book series into the search engine's help desk. There was one thing in particular about the California-based Jeeves that really caught his attention before he signed on as a top executive: whereas the book publisher he worked for, IDG, had $200 million in sales and a stock market value of $250 million in the late 1990s, Ask Jeeves had gone public with just $10 million in sales and $80 million in losses, yet had a stock market value of $5 billion by December 1999. "After a while," Berkowitz recalled, "I wanted to get into the Internet."

Ask Jeeves had a history, to be sure, and while Berkowitz's focus was more on the future than the past, he first set about understanding how a firm once worth billions of dollars could have stumbled, what sort of talent it had on board, and most important, how it was perceived by users. It didn't take long for him to discover that Ask Jeeves, also known as Ask.com, had consistently failed to deliver on its promises, leaving users frustrated. Led on by marketing that suggested they could type questions in plain English into a search box and get accurate answers—a highly appealing proposition, given the complexities of the Internet—millions experimented with Jeeves. Almost without exception, they found the experience lacking. The butler failed to deliver.

Ask Jeeves had gone public in 1999 with a sexy idea but no plan for how to generate profits, making it typical of the era. The company spent millions of dollars—raised in its public offering—on marketing and promotion to build brand awareness and generate traffic on its Web site, all the while failing to invest in the technology needed to support its promises. Its catchy, tongue-in-cheek ads included such questions as "Why is the sky blue?" and "Why can't you find baby pigeons?" In December of 2000, Jeeves's fortunes evaporated, along with those of numerous other dot-com bonus babies. "It went up with the bubble and down with the bubble," Berkowitz said. But he remained a believer in the power of brands, and he recognized that unlike other search engines and Web sites, it had a character, Jeeves, who brought the brand to life. The result was that even after the crash, Ask Jeeves had a strong image and a curiously loyal group of users for a product that didn't measure up.

The executive search firm that recruited Berkowitz to Ask Jeeves presented him as a candidate who understood something about the Internet, content delivery, and running a public company. However, he knew virtually nothing about search, so he spent the first several months on the job in 2001 asking questions and listening to people. "The business was doing $7 million a quarter in revenue. I didn't know how big a mess it was when I took the job," he said. After learning a bit more, he made a decision about personnel that would prove instrumental in reviving the business: he replaced virtually all of the company's senior management, but he did so carefully, one area at a time. "I wanted to make sure I knew how to fix it before I broke it," he said. "You don't change the tires on a car while you are driving."

Berkowitz came to regard search as a distinct product that could have its own brand identity. One of his new hires, Jim Lanzone, recognized that Ask Jeeves had started out as a question-and-answer service, which involved a great deal of manual labor. In contrast, Google had an automated system build around its PageRank formula that could produce faster, more relevant results. Lanzone set out to find a distinctive way for Ask Jeeves to emulate the Google model, which was more technology-driven. At the time he joined the firm in August 2001, Lanzone recalled, although Ask Jeeves had a pretty poor reputation it remained a top 15 Web site. How could this be possible? The answer, Lanzone realized, was that many people who had frustrating experiences on Ask Jeeves kept coming back, hoping things would improve. Neither Berkowitz nor Lanzone took the 15 million to 20 million visitors to the Web site for granted. In fact, they marveled at the power of the brand.

Working together, the two men came to the conclusion that what Ask Jeeves needed to do was buy technology that could deliver answers to people's queries, help the firm retain its user base, and then, over time, if it survived, grow. "We had the audience. Our strategy was to improve the product," Lanzone said. By that point, Wall Street had soured on Ask Jeeves so much that the company actually had more cash than its total stock market value. This was a signal that Wall Street anticipated it would burn through the remainder of its cash and eventually go out of business. Berkowitz and Lanzone had other ideas, and they set about the country looking for acquisition targets. "We needed a search engine," Lanzone said. "That is really the only method you can use to answer billions of queries."

In the fall of 2001, Ask Jeeves set its sights on an obscure, seven-person firm far from Silicon Valley, in Piscataway, New Jersey, called Teoma. (*Teoma* means "expert" in Gaelic, though its name was chosen by a Greek.) Originally funded by the Pentagon and started by some Rutgers University computer science professors, Teoma was a "third-generation search technology," at least in the eyes of Berkowitz and Lanzone. As they saw it, generation one was AltaVista, generation two was Google, and generation three was Teoma, or what Ask Jeeves came to refer to as Expert Rank. Teoma's technology involved mathematical formulas and calculations that went beyond Google's PageRank system, which was based on popularity. In fact, the concept had been cited in the original Stanford University paper written by Sergey Brin and Larry Page as one of the methods that could be used to rank indexed Web sites in response to search requests. "They called their method *global popularity* and they called this method *local popularity*, meaning you look more granularly at the Web and see who the authoritative sources are," Lanzone said. He said Brin and

Page had concluded that local popularity would be exceedingly difficult to execute well, because either it would require too much processing power to do it in real time or it would take too long.

At Rutgers, a 50-year-old Greek computer scientist named Apostolos Gerasoulis, known for his brilliance and for wearing the same T-shirt several days in a row, had found a way to make the Teoma concept work and do local popularity searches in a fraction of a second. But for Berkowitz, persuading the Ask Jeeves board of directors to spend millions of dollars on an East Coast technology after the dot-com crash would not be easy. Jeeves was still not profitable, cash was precious, and the landscape was littered with other once-promising technologies that turned out to be money pits. During a series of board meetings in the summer of 2001, Berkowitz argued that it was more important for Ask Jeeves to buy or build its own search technology than it was to control the process by which it obtained ads. "Google's product is superior to anyone else's," he said. "The key to success is having a great product."

If it had its own search technology, Berkowitz reasoned, Ask Jeeves would maintain control over its users. Some board members disagreed, saying it would be smarter to license search technology developed at someone else's expense. The debate took place against a rather dire backdrop, as the company stock dropped below $1 per share, but Berkowitz ultimately prevailed, raising his hand when a board member asked, "Who is going to take responsibility if it doesn't work?" Very early on the morning of September 11, 2001, he closed the deal he had set his sights on, buying Teoma for $4.5 million. In hindsight this was a bargain, though at the time it amounted to more than 10 percent of Ask Jeeves's stock market value. The deal closed and was announced only a short while before the terrorist hijackings. Instead of expressing delight over the deal or despair over the tragedy, the Greek wizard who had developed Teoma technology began to write. "Apostolos wrote a poem and sent it to the whole company he didn't know yet," Lanzone said. "And then he started hitting a few of us up with, 'How are we going to address the information needs of people coming to Ask Jeeves to find out about September 11th?' He could have been out there counting his money. Instead, he did it for the love of creation and invention."

By December 2001, Ask Jeeves search was running on Teoma technology, and Berkowitz was convinced the firm had a future. "That was the moment of the rebirth of Ask Jeeves," Berkowitz said. "It defined our user experience."

Despite the building blocks of a strong brand and improved search technology, Berkowitz still lacked a way to generate big-time profits. Enter Google. In the spring of 2002, Ask Jeeves's contract for search ads with another firm expired. Berkowitz felt he was now in a position where he could approach Google to see if the company would enter into a new kind of partnership by delivering ads to Ask Jeeves, and then sharing the revenue generated from clicks on the ads.

"The acquisition of Teoma put us in the position to be a free agent," Berkowitz said. "We saw Google as an ad agency. Most people were held hostage to them because they needed Google's algorithmic search. We were able to position ourselves as a unique animal." In addition to an attractive revenue-sharing agreement, Berkowitz also wanted to partner with Google because it was capable of delivering ads to Ask Jeeves users that were relevant to search results, an added benefit. He held a series of meetings with Google officials to explore the prospects of a partnership.

The Google effort was led by its chief executive officer, Eric Schmidt, and its head of global sales, Omid Kordestani. These high-level contacts indicated that the stakes for both parties in a positive outcome were immense. "I had a face-to-face meeting with Eric," said Berkowitz, "and he said how much he wanted this business and how our business base was different from theirs. In a fast-growing market, he said, it was better to work together to grow the market than to kill competitors and shrink the market." Google also allowed Ask Jeeves, before proceeding, to test its ad delivery system on a pilot basis. Berkowitz liked the results, and began to view Google as a partner rather than as the enemy. But the negotiations seemed endless to him; he was used to cutting deals more quickly. Part of the reason was that Larry Page and Sergey Brin had to sign off on every major aspect of the contract as it was being negotiated.

Despite warnings from others in the industry that Google would steal visitors from Jeeves, Berkowitz inked a multiyear deal with the search leader in July of 2002. At the last minute, an additional year was added to the pact so that the announcement, based on annual projections that shared revenue would reach $100 million, would garner more attention. Still, the deal made no sense to many industry watchers, since both Google and Ask Jeeves were search engines seeking users. Berkowitz had a different view. "The idea is you can compete and cooperate at the same time. That is what we were doing with Google. I call it 'coopetition,'" he said. "At the end of the day, Ask Jeeves would only survive by taking bold steps."

The agreement, in fact, had immense upside for both firms. For Ask Jeeves, there was much-needed revenue. For Google, there was proof that it could partner with Web sites and other search engines without their fearing that Google would siphon their traffic.

"This agreement will put our advertisers in front of millions of new users while also offering them new audience characteristics distinct to Ask Jeeves," said Eric Schmidt. "Google and Ask Jeeves have a shared vision of making it easy and efficient for people to search on the Web. By working in partnership, we will help grow the overall search market."

The Google Economy delivered the advertising and revenue Berkowitz had hoped for. Ask Jeeves turned the corner and became profitable by the fourth quarter of 2002. At the same time, it delivered customers to Google it otherwise wouldn't have had. The overlap of users was so small that the partnership made sense for both firms. Google also used the Ask Jeeves deal as a selling point as it sought to sign up more affiliates of all sizes to display its ads and thus join the Google Economy.

As Google added its search and advertising service to the Web version of *The New York Times* and to Amazon.com and other high-traffic sites, the Google Economy built tremendous momentum, increasing the inventory of places where its ads could be displayed, the number of businesses bidding for those places, and the prices they were willing to pay. The Google Economy, in full regalia, also had a self-reinforcing effect: The more computer users who clicked on Google ads, the more money Web site owners would make. The more money they made, the more other sites would be willing and eager to add Google search and ad technology to their offerings. The bigger the network grew, the harder it would be for anyone to challenge it. Just as the major TV networks were the best place to go for advertisers seeking broad audiences, Google was becoming the #1 destination for displaying ads online.

To quantify the power of the Google Economy, it helps to look at the Ask Jeeves financial results, since most of its revenue is derived from ads delivered to it via Google. After posting a loss in 2002, it had revenue of $107 million in 2003, which more than doubled to $261 million in 2004, when Ask Jeeves recorded profits of more than $50 million. The partnership was so successful that the parties later would announce that they were extending it through 2007. Ask Jeeves had become a prime beneficiary of sound management and the Google Economy. "We look forward to maintaining our partnership with Google as we continue to deliver great experiences to our users while growing the immense opportunity in paid search," Berkowitz said.

CHAPTER 12

And on the Fifth Day . . .

Krishna Bharat watched in horror from a hotel room in New Orleans on September 11, 2001, as terrorists attacked New York and Washington. Bharat, a 31-year-old Google software engineer from India, flipped through television channels and furiously searched the Web to learn as much as he could about what had happened. He worried about his family and his travel plans, since all planes had been grounded by the federal government. He quickly forgot about the forum on information retrieval that he was attending in New Orleans, as did others gathered in the hotel with him. Instead, he focused entirely on the worst attack on U.S. soil since Pearl Harbor. Out of such moments of crisis and devastation, great ideas are sometimes born. For Bharat, September 11 marked such a new beginning.

As a young man growing up in India, Bharat was, in his own words, a "news junkie." He read Indian newspapers, watched Indian television, read *Time* magazine, and sat with his grandfather, who lived with his family, listening to radio news reports from the BBC. Recognizing that censorship and cultural issues sometimes got in the way, Bharat soon caught on to the notion that if he really wanted to understand an event, he needed to turn to multiple sources of information, particularly if the news related to India. There were certain subjects that were just too sensitive for the Indian media to treat openly or in full, so each week Bharat eagerly waited for *Time* to arrive, and became fascinated by the notion of how his grandfather kept informed about local and world events. This searing experience as a youth ultimately influenced his thinking about some of what he would do as a grown-up.

After graduating from Georgia Tech with his Ph.D., Bharat moved to California and went to work for Digital Equipment in Palo Alto, where his focus included consulting for the AltaVista search engine. The job increased his interest in Web search, building on his background and education in information retrieval. Along the way, he met Google's founders. "I understood Google technology and AltaVista technology, and I really liked Larry and Sergey and their attitude," Bharat recalled. In 1999 he joined the guys at Google, and with a fellow colleague from Digital he founded Google's research group.

In his new position Bharat worked on a wide array of tasks, since many areas of the search engine had not been developed. Being in the research group also gave him the opportunity to emphasize projects with a longer-term time horizon than those undertaken by most of his Google colleagues. "I worked on things like trying to understand how Google was used and behaves for millions of people," he said.

But Google had something else special about it that Bharat relished: a rule that software engineers spend at least 20 percent of their time, or one day a week, working on whatever projects interested them. The 20 percent rule was a way of encouraging innovation, and both Brin and Page saw this as essential to establishing and maintaining the right culture and creating a place where bright technologists would want to work and be motivated to come up with breakthrough ideas. At other companies, freelancing on side projects and new ideas typically was frowned upon, making it difficult for entrepreneurial employees, who often found they had to work on them in secret, without their boss's knowledge. At Google, the 20 percent approach sent the opposite message—spend one day a week on something you, not your boss, are passionate about, and don't worry about such pedestrian matters as whether the idea could be a moneymaker or something that could be turned into a successful product. In other words, please yourself.

This 20 percent rule was unusual in modern business, but it did have a precedent. Many years earlier, 3M, the company behind the consumer brand Scotch tape, developed a 15 percent rule to spur innovation by directing its engineers to spend a portion of their paid time on projects of their own choosing. For 3M, the extra time to dream yielded, among other things, the idea for Post-it Notes. The more direct influence on Brin and Page to adopt a flex day for its engineers stemmed from the arrangement they had witnessed at universities, where faculty often spend four days a week in the office and pursue research or other projects on the fifth. Since Sergey and Larry had dwelled in a university environment their entire lives until taking leave from Stanford to found Google, extending that collegiate atmosphere of freedom and self-direction to their new company seemed natural.

"The 20 percent time was invented for people to just explore," Bharat said. "People are productive when they are working on things they see as important or they have invented, or are working on something they are passionate about. This is also an opportunity to get bottom-up innovation. There is only so much that top management can specify or ordain." Engineers have the choice of using their 20 percent time each week or pooling it, then spending a month working on a project. "People talk over lunch about the things they are playing with," Bharat said. "It is like they are the CEO of their own little company. Once an idea matures a little bit, they tend to talk about it in a more public forum"—but still only within the confines of the company.

One way of circulating word about what they are doing is through bulletin boards on Google's internal computer network, but the company also sets aside time for peer reviews where engineers can receive feedback on their budding ideas. "Positive feedback implies other people are willing to work with you, and you have the prerequisite then for a project," Bharat said. "Then you can go and build it. Google has a way of having these ideas germinate in 20 percent time and blossom in the ideas forum. Then, some ideas will get funded and become something that management tracks to make sure the project sees the light of day." Rather than having employees moonlighting as inventors at home—with the risk that an idea will either fail from lack of resources or succeed to the point that they quit to pursue it full-time—Google gives them both freedom and resources. "Fundamentally, this is a wonderful paradigm and has contributed to a lot of Google engineers producing things," Bharat said.

When Bharat was a doctoral student at Georgia Tech in the mid-1990s, he indulged his passion for news by developing a new type of newspaper. He had wanted to see if he could devise a way for a crawler, an online vacuum cleaner of sorts, to take similar wire service news stories from disparate Web sites and bring them to a single place, where they could be repackaged and easily read by people interested in those subjects. "This was my first taste of online news," he said. "There were not many online news sources at the time." In addition, Bharat, who had a sharp eye and a knack for organizing information in easily accessible ways, wanted to find a way to design a newspaper that catered to the interests of individual readers and their habits, rather than coming up with a single design for everyone. "My original idea was really better layout. I know how to customize things nicely. The way news was organized in a newspaper format online, you had to click on a link and then come back. I said, 'I can do a better layout than that.' Then I realized we could observe what people do. So maybe we could turn this around and make tomorrow's newspaper, ahead of time, to your liking."

It all came together on September 11—Bharat's upbringing in India surrounded by multiple news sources; the work he had done on newspapers at Georgia Tech; and Google's emphasis on 20 percent time and innovation. On a day when access to credible information really counted, he began conceiving of a better, faster, and easier way for computer users generally, and journalists specifically, to learn what was being written and said around the world. In the back of his mind lay the memory of his Indian grandfather listening to the BBC. "There was so much going on and so many points of view: the American view, the world view, the Afghanistan view, the European view. It was so fascinating. I realized the Web was inconvenient for researching a topic you wanted to understand broadly," Bharat said. "Every newspaper did a great job of putting out content, but they didn't have the time, or inclination, to cross-link to other subjects. News is also real-time. There was no way for a *Washington Post* reporter to go out of his way to find other articles being written on the same subject. Of course, Web search engines then were not making it easy for you to find other relevant articles. Readers like me, and journalists themselves, could benefit from this, since it took a long time to find out what others had written or said. Especially on a topic like September 11, with so many differing opinions, I decided this was a problem worth solving."

For the next few months, Bharat worked on solving the problems he had seen so vividly on September 11. There were so many issues and subsidiary issues to tackle at once. Fundamental to his vision was an exploration of whether he could write a set of mathematical equations that would simulate, on an automated basis, the kind of decisions that experienced newspaper editors made about stories and layout. Using a technique known as clustering, he began dividing stories into categories, ranging from world news to politics to business to sports, and then watched how little or much editorial activity was generated by a specific story. Next, he began adjusting the rank of stories based on their origin, with greater weight given to stories written by reporters for leading U.S. newspapers and wire services, including *The New York Times*, *The Washington Post*, the Associated Press, and Reuters. At the same time, inclusiveness was important, so no matter how big and prestigious or small and obscure a news source might be, Bharat wanted to find a way to include it.

Given the importance of updates to the news and the real-time nature of what he envisioned, Bharat's formula also increased the rank of fresher stories over older news. "This must be computed over and over again. This became a real-time operation," he said. In creating different editions of his online news site, he also added relevance to the mix. For example, all other things being equal, a U.S. story would be of greater interest to computer users in the United States than a Canadian story, and vice versa. This might be of particular value if different editions of Bharat's project, dubbed Google News, came into being.

By late 2001 or early 2002 he had developed the initial version of StoryRank, a first cousin of the PageRank formula used to prioritize Google search results. But it was not sufficient, he realized, to show headlines alone, so with the help of others he built a search function specifically for the online news he was collecting. There would always be much more news than could be shown on any homepage, but a search for the news would let the user decide, in a Google-like way, what subjects to highlight.

Two other Googlers worked with Bharat to build a working demo, which was released within the company to gauge reaction from others. He knew he was onto something big in December 2001, when CEO Eric Schmidt dropped in to chat about Google News. "I knew this was popular," Bharat said. "I got a lot of feedback from engineers. Suddenly Eric Schmidt walked into my office and said, 'This is a cool product.' I don't even know how long he had been at Google at that time, but he was really interested and wanted to make it happen. Then, subsequently, I spoke to Larry and Sergey, and they were both very interested."

With excitement in the ranks and the endorsement from the top, Bharat's 20-percent-time project turned into a full-time endeavor. For Bharat, it was the fulfillment of a dream. And in keeping with Google practices, he received the resources necessary to take the demo and build it into something that would shine online for millions of users. That meant designers who would focus on the user interface; a seasoned product manager, Marissa Mayer, to shape the product and study it from the user perspective; and a team of engineers to refine and test the software that crawled the Web, ranked stories, and organized disparate information into a comprehensible whole. "At Google, if something is worth doing, it gets funded," Bharat said, noting that no one ever asked how the product would make money.

What gave Google the right to republish news produced by various media companies on its own Web site? Nothing, really. But the concept caught on so quickly that news organizations wanted to be a part of it. Google News always identified the origin of the news stories it carried and let users click to be taken to that source. Google, in effect, was serving as a news broker. It didn't pretend to own the news it was republishing, which meant the company did not necessarily need to license and pay for the news it retrieved from hundreds, and later thousands, of media outlets. "I much prefer this model," Bharat said. "We want to be the information access point, but we don't want to own the content. We do what we are best at, which is bringing people to content that matters to them quickly."

Unlike the clean, uncluttered Google homepage, the Google News page was packed with headlines and content. "We wanted to make sure the information was really dense. It had to be a content-rich page to get a lot of information to the user," said Mayer. Using newspaper lingo, she said that in selecting a final design, much thought also was given to the quantity of information a user could see "above the fold," the term that referred to the top half of a newspaper's front page. In this instance, "above the fold" referred to stories that users could see without scrolling.

Google News caught on with computer users and journalists alike, leading to new innovations such as Google Alerts, an automatic way for people to track specific topics of interest by email. (Bharat did not develop the alerts. By that point in time, he had moved on to head Google's research and development office in India.) Millions signed up to receive the alerts, which proved invaluable to people following a particular company, issue, individual, or subject in the news. For journalists once fearful of missing a story, Google Alerts, in combination with the Google News homepage and search function, made covering a beat more efficient. It also led to a greater sharing of ideas, since articles from an array of sources around the world—from the leading metropolitan dailies to small-town tabloids—would be read more widely.

"The very fact that news articles on various subjects are juxtaposed increases news reading," Bharat said. "Google News provides the opportunity to find out information, and then I want feedback from people who read the information. I want there to be a bigger debate about issues. I would love to see that happen."

Craig Nevill-Manning joined Google in 2001 after working as a professor at Rutgers University in New Jersey, where he had focused on finding innovative ways to extract information from Web sites. A native of New Zealand, he was attracted to Google not only by the idea of doing interesting research, which he was doing already, but also by having his ideas turned into products that people could use. Born in May 1969, he was older than most Googlers, having entered the world two months before astronaut Neil Armstrong became the first man to walk on the moon. Given his age and stature, Nevill-Manning, who had done a postdoctoral fellowship at Stanford, carried the title of senior research scientist. Like others, he was attracted to Google by its culture of innovation, including 20 percent time. It didn't take long after his arrival at Google for Nevill-Manning to use his time to begin exploring the field of online retailing.

"I was talking with another product manager about the fact that when people came to Google looking to buy, we didn't necessarily do a good job of helping them. We didn't have good coverage of what was on the Web, and you often want to be able to look at comparative products and prices. We started to think about what we could do to make this better," he said.

Nevill-Manning studied various online retailers closely, including Amazon.com, looking at how they extracted information based on category, price, and description. After about six months, he turned his idea into a prototype. On a lark, he named the computer file where he stored the code for his new creation "Froogle," since it rhymed with Google and conveyed the hunt for value that consumed consumers. Internally, the official code name for the initiative was "Product Search." In early 2002 he took his prototype to Larry, Sergey, and Eric to get their reaction. They liked the idea, but weren't sure whether it was worth further exploration. One lingering question was whether Froogle could become part of Google's core search or had to be its own product offering. Nevill-Manning recognized that the right thing to do was focus on his day-to-day work on Google's advertising system, and he set the idea aside. "We have a million good ideas at Google and can only devote significant resources to a small subset of them. To do this would mean several engineers and taking months to do it. I shelved Froogle for a while and went back to work on AdWords," he said.

But as time went by, Nevill-Manning found it hard to let go of Froogle. He began refining the concept and persuading himself, and eventually others, that it was worth pursuing aggressively. The guys at the top were not so sure. "It sort of hung there while people thought about it and how it would fit," he recalled. "We got clearer on what the product was. We generate many, many ideas through this 20 percent time and then there is a funnel because we can't take all of these ideas and make them products. Froogle was trying to get through that funnel." Every so often, he would walk into Sergey's office to talk about Froogle and show him the latest features and ideas. The hurdle that he had to overcome was whether Froogle could be comprehensive and scalable, like the core Google search engine.

"The proof of the pudding is having Larry say, 'I am looking to buy a new digital camera,' and he typed that in. Or Sergey saying, 'I know of some green laser pointers that just became available. Let's see if we can find them on Froogle.' Fundamentally it came down to, 'Can you find those new or obscure products?' Those are the things Google is good at finding in the nooks and crannies of the Web. That is what it all came down to. How useful was it for Larry and Sergey?"

Eventually, Brin relented, giving Nevill-Manning the green light to staff up by putting together a fresh team of engineers to build Froogle. The challenge in writing the code was unlike many other Google projects, since it did not fit in the mode of PageRank. Products available on the Web typically did not have links indicating popularity, giving rise to the need to figure out a different way to determine the most useful and relevant product results from a Froogle search. The race was on to get Froogle up and running in time for the busy holiday shopping season in 2002, but snafus pushed the product debut back to mid-December, and the late arrival meant its initial impact was minimal. By the next holiday shopping season, in 2003, Froogle had gained an audience, deriving most of its traffic from Google users who had tried it out. In the Google tradition, Froogle distinguished itself from many other shopping sites by refusing to accept paid listings in its main search results, opting instead for small, clearly marked text ads that appeared to the right of the product listings. "Users like to see everything available in one-stop shopping," Nevill-Manning said.

A debate about 20 percent time at Google, and whether it is substantive or window dressing to attract technologists, ignited among Googlers and Microsoft employees, adding to the tension between the firms. Google engineer Joe Beda praised 20 percent time and explained why it worked well inside Google headquarters. "There is a big difference between pet projects being permitted and being encouraged," Beda said. "At Google, it is actively encouraged for engineers to do a 20 percent project. This isn't a matter of doing something in your spare time, but more of actively making time for it. Heck, I don't have a good 20 percent project yet and I need one. If I don't come up with something, I'm sure it could negatively impact my review.

"The intrapersonal environment at Google is very energizing. When someone comes up with a new idea, the most common response is excitement and a brainstorming session. Politics and who owns what rarely enter into it. I don't think that I've seen anyone really raise their voice and get into a huge knock-down-drag-out fight since coming to Google," Beda said.

"Can 20 percent time work at other companies? I'm sure there are going to be others that try. However, I think that it is important to realize that it is a result of an environment and a philosophy; I don't think that it is something that can be imposed in an independent way. I'd like to stress that these comments are my own and aren't any sort of official word from Google. Please don't draw any grand conclusions of corporate strategy and such."

But a Microsoft technologist has a different point of view. Google, he claims, may not be as different from other technology firms as the company, or its employees, like to pretend. "Why doesn't Microsoft have 20 percent time? Well, speaking for myself, if Bill Gates told me I could have 20 percent time I'd say, "That's nice, Bill, but I'm already working on what I want to work on.' Maybe I'm a rarity at Microsoft, but I don't think so. If any employee isn't happy where he or she is, there are lots of opportunities to move around. The thing is, I need 20 percent time to do all the work I hate to do. Like expense reports or other corporate procedures and processes. Now if someone could take those kinds of chores away, then THAT would be an employee benefit I'd really like."

CHAPTER 13
Global Gooooogling

Meng-Ying Lee, a 26-year-old native of Taiwan, is the epitome of a global Googler. He is one of a new breed of computer user who accesses the world through search, juggling languages and time zones to keep tabs on news, business interests, and family and friends scattered across multiple continents. Polite, animated, and driven, Lee was a latecomer to computers, but he caught on quickly, thanks in part to Google's ease of use and clarity. Like millions of others who surfed the Web for the first time in this new millennium, he has no concept of an Internet without Google. When he arrived in the United States in 2001, fresh from two years of mandatory service in the Taiwanese army, he knew just a smattering of English and nothing at all about how to use a computer. Through an exchange program with Taiwan, Lee enrolled at Webber International University in Florida, where he studied hospitality management and, even more important, learned his way around a desktop PC.

Now an M.B.A. student in northern Virginia, Lee uses Google in English by day to do research for school projects and at night stays up until 2 or 3 A.M. googling in Chinese on his home high-speed Internet connection. He finds that searching the Web in Chinese, his native language, is easier and more productive for familiar, everyday tasks. For him, these include checking his stocks; reading or watching news about America, China, and Taiwan; and following the NBA scores, on which he might have a little money riding. He does all this through Chinese language search. But with so much of the Web still English-only, Lee also looks for English pages and uses Google's automated page translation tools as a guide when he's stuck. Sometimes he dumps unknown words into Google's side-by-side translator to get at least a rough sense of meaning, and perhaps learn some new phrases along the way. He sets Google as his homepage, and uses a Google-powered site in Taiwan for Chinese-language surfing. (At the sushi restaurant where he works, delivery runners also use Google on the computer in the back room to check addresses and directions before heading out.)

For a young man of modest means who until quite recently lived unconnected to the global high-tech economy, Lee now has a wealth of information at his fingertips, fed in large part by Google. The Internet and e-commerce revolution that he missed in the 1990s can now be his as a second generation of the Web comes of age. He talks excitedly about university friends from Sweden and Bulgaria who have already launched international businesses; he is eager to do the same. And while his parents now live in California, Lee's own business aspirations lie in Taiwan and China, which boasts a huge marketplace that is waking up to the Web. By emailing with friends and googling for information, he is sizing up prospects for side ventures as he simultaneously pursues a career in hospitality. He is moving fast because he sees great opportunities and knows that he will have competition—and that they too will have Google.

By 2003, tens of millions of people daily were searching Google in their native tongues, choosing from a list of nearly a hundred available languages. In Greek, Latin, Gaelic, Hindi, Ukrainian, Urdu, Croatian, Czech, Esperanto, Persian, Portuguese, Norwegian, Swedish, Spanish, Swahili, Thai, Malay, Afrikaans, Maltese, Chinese, Japanese, Tagalog, Basque, Icelandic, Italian, Indonesian,

Dutch, Danish, Zulu, Korean, Welsh, German, French, Arabic, Hebrew, Latvian, Lithuanian, Romanian, Slovenian, Russian, Finnish, and English—and for fun, Pig Latin, Klingon, Elmer Fudd, and the Swedish Chef's "Bork, Bork, Bork!"—they looked on Google for everything from the basic necessities of food and shelter to commerce, education, leisurely pursuits, and of course, sex.

Around the world, businessmen, investors, and their lawyers would consider themselves fools to do a deal without googling the other party. Authors writing books, including this one, find facts and files fast using Google. High-ranking government officials use it to find documents themselves instead of asking aides to track them down. When scientists attack certain vexing problems, they google the genetic code to discover relationships they didn't know existed. Teenagers eager to know the words to a popular song simply google it. Accomplished chefs and hungry hacks with leftovers in the refrigerator google the ingredients to figure out what meal to whip up quickly. CIA agents use Google to track terrorist groups. Software engineers turn to Google to answer computing questions instead of opening a book or asking a colleague. Patients google their ailments. Employees google their bosses. Athletes google their competition. And globe-trotters and armchair travelers alike use Google to learn about far-off destinations without ever leaving home.

"Recently, Google has helped me plan my wedding in Chile," said adventure traveler Erica Smith of Bethlehem, Pennsylvania, one of many Google users interviewed for this book. "Between a good guidebook and a good search engine, I've done all the planning from my sofa. I'm not using a travel agency and I've never been to Chile—it's happening because I'm able to access the right information." Matt Stedina, a Vermont builder and fishing guide, turns to Google to investigate his competition for upcoming flyfishing tournaments. "I like to know what part of the country these guys are from. I want to know what lodge they work for, and how many guides they really do," Stedina said. "Google has helped even me, a low-skill computer guy. It really has made it a small world."

Private hedge fund manager Mark Cordover was impressed with how handy the sponsored links on Google were during a home improvement project. "I needed stainless steel bolts for a deck on my house. I went to Home Depot and came home two hours later empty-handed," Cordover said. Searching on Google.com, he found 12 vendors for the elusive bolts. "Had any of these vendors bought 30-second ads on the Super Bowl, I couldn't have cared less, but when I needed my stainless steel bolts, they are paying Google for the opportunity to be presented to satisfy a need." Michael Sladek, a manager at the arcade maker Skee-Ball, was drawn to Google by "the incredible free pop-up blocker" on its browser toolbar. He stayed with Google because of the way it provides natural search results front-and-center, and targeted ads on the side. "They give you the answer first; then you can look at all of the 'Web deals' if you want to," Sladek said. "They are less invasive."

According to *Wired* magazine, a special category exists among the rich and famous, called "Google SuperUsers." While they may google themselves or google to stay abreast of the news, they also have interesting ways of searching to help them work and play. Gary Trudeau, the cartoonist who created "Doonesbury," googles as he draws. "Google is my rapid-response research assistant. On the run-up to a deadline, I may use it to check the spelling of a foreign name, to acquire an image of a particular piece of military hardware, to find the exact quote of a public figure, check a stat, translate a phrase or research the background of a particular corporation. It is the Swiss Army knife of information retrieval." Trudeau is not the only creative talent who makes regular use of Google. John Gaeta, visual effects supervisor of The Matrix trilogy, revealed to *Wired* that he too is a devotee. "Within the last seven days, Google has altered and augmented my perceptions of tulips, mind control, Japanese platform shoes, violent African dictatorships, 3-D high-definition wallpaper, spicy chicken dishes, tiled hot tubs, biological image-processing schemes, Chihuahua hygiene, and many more critical topics. Clearly, thanks to Google, I am not the man I was seven days ago."

Other Google features, including Google News, are hugely popular, especially in a town like Washington DC that is populated by news junkies. "I can't imagine life without Google News," said Michael Powell, then chairman of the Federal Communications Commission. "Thousands of sources from around the world ensure anyone with an Internet connection can stay informed. The diversity of viewpoints available is staggering." Wes Boyd, president of MoveOn.org, which promotes progressive political views in Washington, said, "Google rocks. It raised my perceived IQ by at least 20 points. I can pull a reference or a quote in seconds, and I can figure out who I'm talking to and what they're known for—a key feature for those of us who are name-memory challenged."

Michael Chabon, author of *The Amazing Adventures of Kavalier & Clay*, said, "Writers of the past had absinthe, whiskey or heroin. I have Google. I go there intending to stay five minutes and next thing I know, I've written 43 words, and all I have to show for it is that I know the titles of every episode of 'Nanny and the Professor.'" Matt Groening, creator and executive producer of *The Simpsons,* finds Google indispensable. "It's not my homepage, but it might as well be," he says. "I use it to ego-surf. I use it to read the news. Anytime I want to find out anything, I use it."

"I can't explain it—it's just a funny feeling that I'm being Googled."

No other brand has achieved global recognition faster than Google. The company's name has entered the lexicon not only in English but in several other languages too; Germans *googelte*, Finns *googlata*, and the Japanese *guguru*. But not everybody has easy access to Google. Surfing the Web in many parts of West Africa, for example, is still painfully slow and expensive, dependent on aging computers and dial-up connections. Many people, especially younger generations, are aware of Google and eager to use it for information and self-betterment, but access and infrastructure cannot keep up with demand. In war-torn Liberia, Prince Charles Johnson III, a college-educated driver for a U.N. mission, was able to use Google for most of his class assignments in economics and management while in school. Now he likes to keep up with news about American politics and President Bush. "I love this guy," Johnson writes. "It was because of Google that I know the entire First Family, Laura, Barbara, Jenna, Barney, Miss Beazley [dogs] and Willie [cat]." But while Johnson can access the Internet at work, the standard rate of $2 dollars per hour for logging on at Internet cafés is a luxury for all but the wealthiest Liberians. In neighboring Guinea, however, access is cheaper, although still slow and intermittent because of electricity woes. At the Cyber Ratoma, an Internet café in the capital city, French-speaking Guineans use Google.fr to search online educational opportunities, business leads, and medicines. Like Meng-Ying Lee, searchers there find that many Web resources tend to be in English, but they first turn to someone nearby for help, such as café proprietor Diallo Mamadou Sarifou, rather than Google's automated service. Sarifou says that Google's speed and simple interface make the laborious process of Web surfing in the developing world a little easier to bear.

What jet travel began in the 1960s, and cheap international phone calls and email have accelerated in recent years, has now been pushed to a new realm by Google and its search engine kin: the elimination of geography as a barrier to communications and commerce. From home or office, people can make meaningful contact with strangers on the other side of the planet and google them to learn their personal history, see what they look like through Google Image Search, find their phone numbers and Web sites, and view satellite images of their homes.

But the practice of googling raises important questions of privacy and etiquette that have yet to be widely or definitively addressed. Just as cell phones and email brought new codes of contact that had to be learned over time, Google has forced people to confront and agree on new behaviors. What is the distinction between the harmless googling of an individual and cyberstalking? Should people tell each other up front that they've googled them, or pretend not to know the details they learned in a search? In a world where people's pasts are increasingly laid bare on the Web and even a cursory search can uncover uncomfortable or embarrassing details, these are questions without easy answers.

The disturbing reality is that the Internet is replete with out-of-date, conflicting, and inaccurate information. Rumor mills abound, and even trustworthy sites can be slow at updating facts and figures, leaving both googler and googlee exasperated. A greater injustice in the eyes of some are the old or unflattering photographs that make it into Google's Image Search. It is difficult if not impossible to get such things removed—or to chase down a trail of negative or false information once it's been able to fester on the Web. Just ask Sergey Brin, who has several photos from his Stanford days—including one of him dressed in drag—still floating around the Web.

To the delight of cybersleuths and the chagrin of the hunted, old Web pages are granted a life after death by Google's practice of storing, or caching, a copy of every page it downloads. Thus, even after a page has been taken down by its author, a Google searcher can find and retrieve it. A Web site called the Wayback Machine, hosted by the nonprofit group Internet Archive, goes even further; they've amassed an impressively thorough clickable history of the Web going back to 1996 that turns up in some Google searches. Early versions of the Google site, for example, and Page's and

Brin's personal Stanford homepages, live on in the Wayback Machine.

It is easy to make light of the dangers of promiscuous googling, yet important rights and personal safeguards do come into play. Job hunters or coop applicants wishing to expunge one-time indiscretions or criminal pasts may be at the mercy of a googler. As Google's reach and influence continues to grow, the snooping it facilitates will bring closer government scrutiny and perhaps new laws to govern digital detective work.

A *New Yorker* cartoon that ran in late 2002 amusingly captured the shifting times: over drinks in a bar, one man says to another, "I can't explain it—it's just a funny feeling that I'm being Googled."

Students of all ages are heavy Google users, although some teachers and professors actively encourage the use of more specialized academic search engines, and also encourage the use of libraries, face-to-face meetings, and other time-tested, traditional ways of extracting important information. Educators remain divided over the merits of Google. Many say it makes students lazier, encourages plagiarism, and hinders the learning process by encouraging one-touch rapid research, rather than diligent digging that is driven by a desire to know more about a subject. Others praise it, saying its ease of use encourages exploration of primary documents and analysis at all hours of day or night. They also argue that it minimizes the differences students face, whether their school or university is large or small, whether they are rich or poor, and whether they have access to a terrific library or none at all. In short, they support Google's goal of democratizing access to information, including an increasing amount of scholarly research.

Whatever opinions one has about the academic debate over Google, students find it useful for both learning and procrastinating. Daniel Sabido, a sophomore at the University of Pennsylvania, wrote:

> I have used Google ever since beta and almost every week I learn something new. Now I have started using it as a calculator, since it turns out that if you type something like "3+4–sqrt(16)," Google will compute that for you. Also, it is extremely useful for cooking, especially to a metric-system-raised international student like me. If you type "4 gallons in pints," it will do the conversion for you.

Laura Counihan, an undergraduate business student at Penn, wrote:

> The world is full of tiny questions. Are swans really vicious animals? Did Jennifer Lopez get a nose job at the start of her career? What does it mean to say that "Chivalry is dead"? Google has the answers. I use it many times/day.

Penn student and employee Joanne Murray wrote:

> Every year our department needs to report on the current jobs of our graduates. Our alumni database only knows what students report (and many hide from alumni and development folks!). A Google search is more accurate and quicker than an alumni database search.

Peter Fader, a marketing professor at the University of Pennsylvania's Wharton School, considers himself a "Google zealot," but he sees a real danger in his students relying on it too much. "It is so good it gives you the impression that everything worth covering is covered," Fader says. "That is not true, especially when you are in serious research mode." Much of the Internet is locked away, and it is important to recognize the occasions when another resource, like a specialized database, will give better results, he says. "I will trust Google implicitly for anything it does, but I don't ask it to make toast for me."

What Fader does ask it to do, on an almost daily basis, is find images on the Web to help bring to life the marketing examples he uses in class. "Any given lecture, I probably have six or seven bits in there that came from Google Image Search," Fader says. By speeding up the tracking of research citations, Google also comes in handy during decisions about whether to hire or award tenure to faculty members. Especially for new research, Fader says, "you get a much earlier read on its overall impact and ability to hop across academic disciplines through Google."

In the vast universe of Google users, a small cadre of aficionados and evangelists exists who closely track every move of the search industry leader and act as a kind of editorial board and echo chamber for all things Google. Most of the men—they are almost exclusively men—hold day jobs and have no grand journalistic aspirations, just a deep interest in the technology of search and a Web site to share their thoughts.

Philipp Lenssen, a 28-year-old German programmer, was on an extended holiday in Kuching, Malaysia, when he thrust himself into the mix of professional Googlewatchers. Stuck "doing small favors for friends, but without a real job" because of work permit issues, he passed the time in Malaysian Internet cafés, working on personal projects and delving into technical articles about search engines. When he began posting Google-related entries to his online diary, or blog—writing the sort of stuff he wanted to read but couldn't find anywhere—he noticed that other people were reading it too. Lenssen called his diary "Googlosophy Blogoscoped," a *Monty Python*–esque name he thought would convey the pseudoscientific but fun tone he hoped to take. But not wanting to run afoul of Google trademark policies, which frown on "Google"-stemmed words, he shortened the name to Google Blogoscoped.

The unusual handle helped Lenssen's site stand out in the crowd of blogs cropping up all over the Web. Taking their name from a contraction of "Web log," blogs were a new kind of personal homepage, characterized by spontaneous, time-stamped diary entries on all manner of topics, from gardening to politics to travel. Most were just small-time hobbies, but a few A-list blogs were

Larry and Sergey leased space in a Menlo Park garage for their first office.

An early Google computer, partially constructed from imitation LEGO blocks.

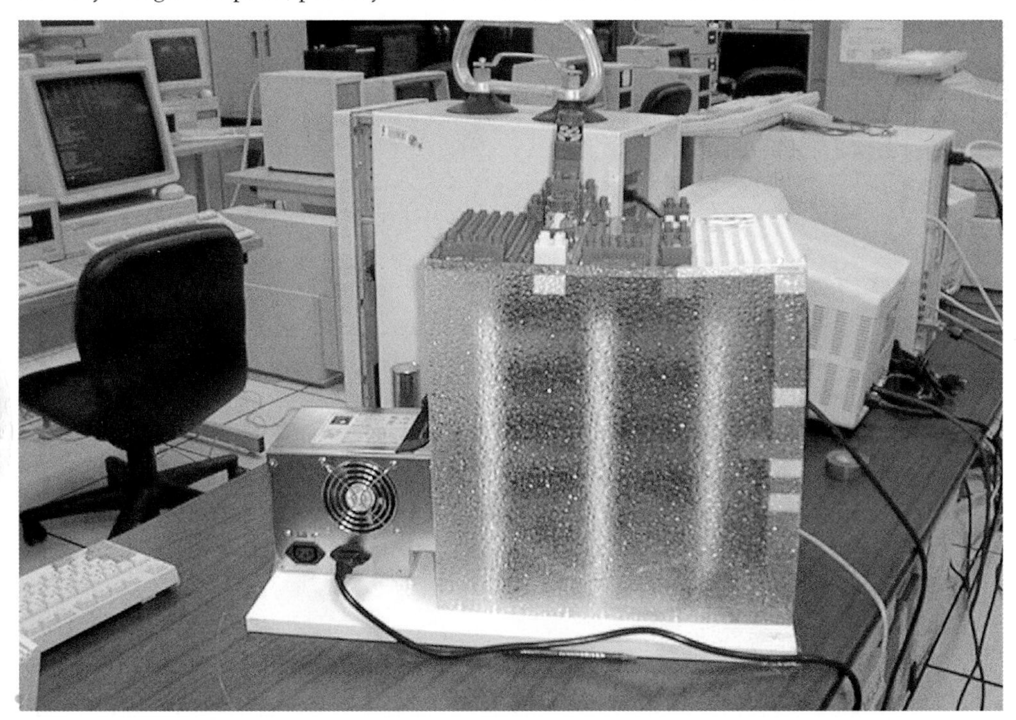

John Doerr of the venture capital firm Kleiner Perkins, which jointly with Sequoia Capital invested $25 million in Google.

Michael Moritz of Sequoia Capital. The two investors threatened to pull out when Sergey and Larry were slow to hire a CEO.

Employee #1, Craig Silverstein, and Google's first female engineer, Marissa Mayer, who led efforts to test and improve the homepage.

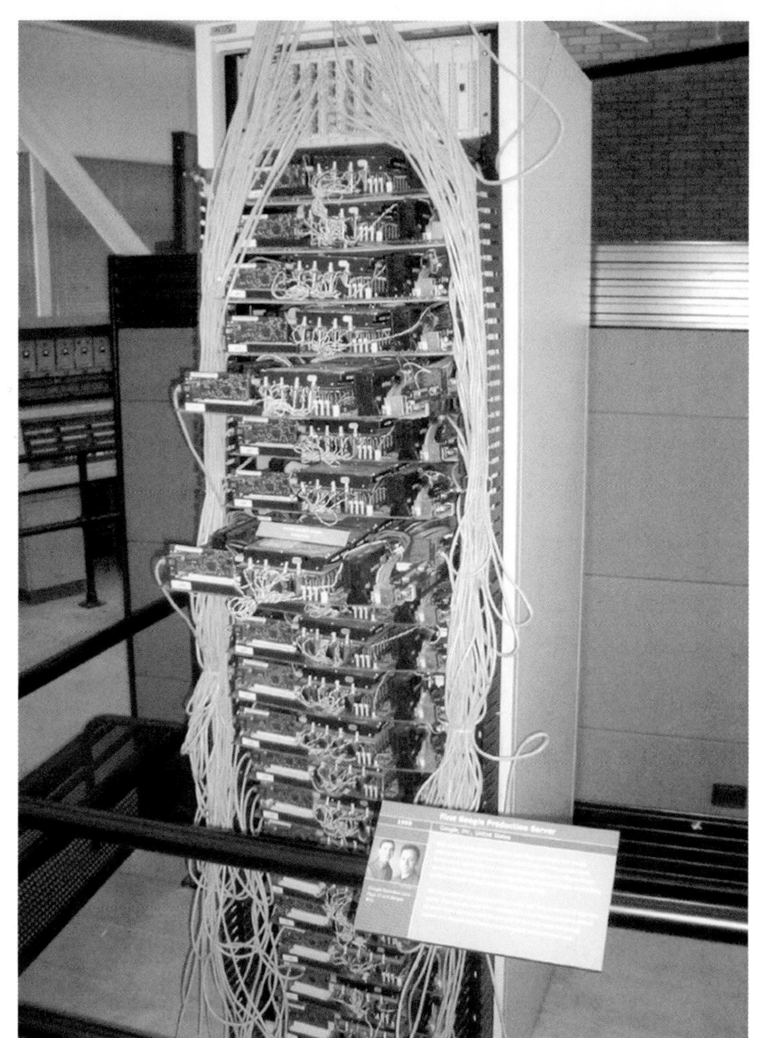

Google's immense computing power comes from ordinary PCs stacked in tall racks and linked by cables and custom software. This early server now sits in the Computer History Museum.

Beloved Google chef Charlie Ayers, who prepared healthy and delicious meals and fostered a fun-loving culture for the company.

As the Google Economy expands in 2003, search impresario Danny Sullivan and Sergey have a "fireside chat" at a jam-packed industry conference.

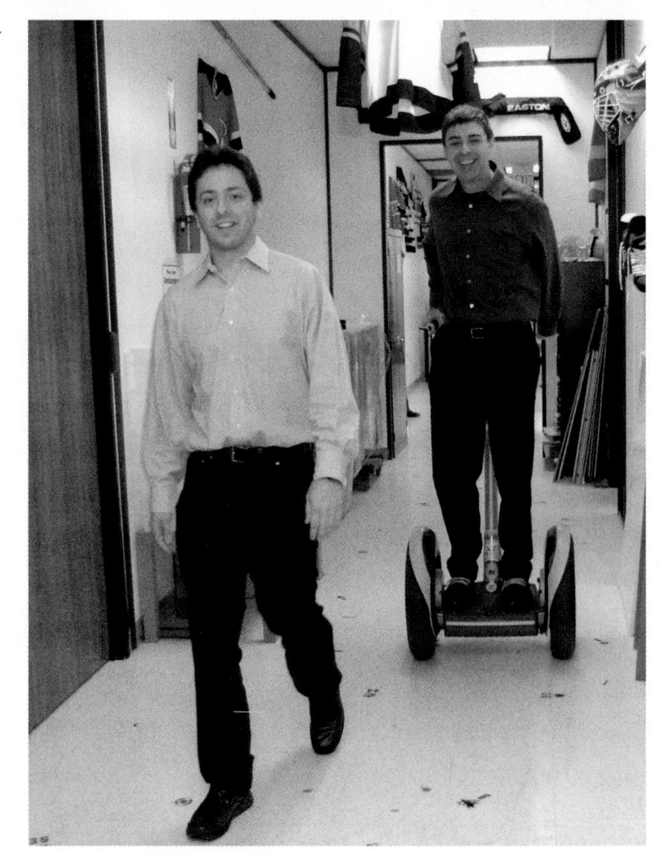

The founders were serious about having fun at the Googleplex, indulging in food, gadgets, sports, and parties.

The space-age Googleplex campus in Mountain View, California, as seen from a B-24 bomber.

Google's global reach extends to a corner in Cairo, Egypt, where an Internet café uses the logo to attract customers.

Software engineer Matt Cutts (center), aka "Porn Cookie Guy," takes questions from Web publishers eager to have their sites ranked higher in Google search results.

In the hot tub at their Menlo Park office, Sergey and Larry could work and relax at the same time.

The future of Google will include work on genetics with maverick scientist Craig Venter, who mapped the human genome.

An interview with Sergey and Larry in Playboy magazine, published days before Google's IPO, sparked a probe into whether it violated stock market rules.

PLAYBOY INTERVIEW: **GOOGLE GUYS**

A candid conversation with America's newest billionaires about their oddball company, how they tamed the web and why their motto is "Don't be evil"

Just five years ago a googol was an obscure, unimaginable concept: the number one followed by 100 zeros. Now respelled and capitalized, Google is an essential part of online life. From American cities to remote Chinese villages, more than 65 million people use the Internet search engine each day. It helps them find everything from the arcane to the essential, and Google has become a verb, as in, "I Googled your name on the Internet and, uh, no thanks, I'm not interested in going out Friday night."

In addition to being the gold standard of Internet search engines, Google is setting a new example for business. It's difficult to imagine Enron or WorldCom with a creed similar to Google's: "Don't be evil," a motto the company claims to take seriously.

This maxim was perhaps most apparent in May when the company announced it was going public. Google founders Sergey Brin and Larry Page explained their lofty ambitions. "Searching and organizing all the world's information is an unusually important task that should be carried out by a company that is trustworthy and interested in the public good," they wrote in an unprecedented letter to Wall Street. With the release of the letter, Newsweek reported, "The century's most anticipated IPO was on, and the document, revealing the search giant's financial details, business strategy and risk factors, instantly eclipsed Bob Woodward's Iraq book as the most talked about tome in the nation."

Page, 31, is the son of Carl Page, a pioneer in computer science and artificial intelligence at the University of Michigan. Larry was surrounded by computers when he was growing up and once built a programmable ink-jet printer out of Legos. Reticent but wide-eyed and reflective, he is Google's clean-cut geek in chief, the brilliant engineer and mathematician who oversees the writing of the complex algorithms and computer programs behind the search engine. His partner, Brin, 30, is a native of Moscow, where his father was a math professor. As Jews, the Brins were discriminated against and taunted when they walked down the street. "I was worried that my children would face the same discrimination if we stayed there," his father told Reuters. "Sometimes the love for one's country is not mutual." The family emigrated to the U.S. when Brin was six. A part-time trapeze artist, Brin is the company's earnest and impassioned visionary—a quieter, nerdier Steve Jobs. Early on, when Google CEO Eric Schmidt was asked how the company determines what exactly is and is not evil, he answered, "Evil is whatever Sergey says is evil."

Page and Brin met as graduate students at Stanford University. After years of analyzing the mathematics, the computer science and the psychological intricacies involved in searching for useful information on the ever-growing World Wide Web, they came up with the Google search engine in 1998. It was far superior to existing engines, and many companies, including Yahoo and MSN, licensed it. (Yahoo recently severed its ties with Google, introducing its own search engine. Bill Gates, who once admitted that "Google kicked our butts" on search-engine technology, has announced that Microsoft will launch its own search engine next year.) With its simple design and unobtrusive ads, Google has quickly become one of the most frequented websites on the Internet, and the company is one of the fastest growing in history. The financial press has estimated that after the initial public offering, Google will be valued at $30 billion, and Brin and Page, each of whom owns about 15 percent, will be worth more than $4 billion apiece.

The two are unlikely billionaires. They seem uninterested in the accoutrements of wealth. Both drive Priuses, Toyota's hybrid gas-and-electric car. It is impossible to imagine them in Brioni suits. Brin often wears a T-shirt and shorts. Page usually dresses in nondescript short-sleeve collared shirts. Both rent modest apartments. Their only indulgences so far fall into the realm of technology, such as Brin's Segway Human Transporter, which he occasionally rides around the Googleplex, the company's Silicon Valley headquarters. (Page often scoots around on Rollerblades or rides a bike.) Page bought a digital communicator that employs voice-recognition technology to place phone calls. Both men are notorious workaholics, though The Wall Street Journal, which uncharacteristi-

LARRY PAGE: "People were checking out who they were dating by Googling them. I think it's a tremendous responsibility. You have to take that very seriously."

SERGEY BRIN: "Any web mail service will scan your e-mail. It scans it in order to show it to you. We are very up-front about it. That's an important principle of ours."

PAGE: "The amazing thing is that we're part of people's daily lives, just like brushing their teeth. It's just something people do. It's quite remarkable."

PHOTOGRAPHY BY ©KIM KULISH

BRIN: "The solution isn't to limit the information you receive. Ultimately you want to have the entire world's knowledge connected directly to your mind."

Larry Page presides over the opening of the NASDAQ stock exchange on August 19, 2004, as Google goes public.

Larry and Sergey, billionaires by age 31, continue to aggressively recruit and innovate, while CEO Eric Schmidt (center) pilots Google's day-to-day business.

drawing significant readerships and revenue. By displaying ads from Google's affiliate network, popular blogs were earning sizable income from paid messages specifically targeted to the content on the page.

As Lenssen's traffic grew, he jumped in the advertising game as well, signing up for the Google network and soliciting some additional sponsorships on his own. At first, the money coming in was just a trickle; but over time, his various Web sites began to earn a small profit. By eliminating the need for an ad sales staff and the high-minimum contracts that traditionally kept homegrown businesses at home, Google had created an entire new economy of small, independent publishers.

"I'm happy not only for the cash [from Google], but for the actual related links it provides," Lenssen says. Just as Armani and DKNY ads boost the allure of glossy fashion magazines, he finds that Google's text ads enhance—in a more targeted way—the content of his often threadbare site. The thousands of people who read Google Blogoscoped come from all over the world, he says. He posts personal commentary and analysis, news items, interviews with industry players, and, every so often, a breaking story that gets picked up by other bloggers. His biggest scoop involved a Google employee who candidly wrote about internal matters on his own blog and was summarily fired. "It was not exactly Google-gate here," Lenssen says, "but mainstream news picked this up and it created quite a stir."

When Google launched its own official blog, Lenssen's site and about two dozen others were listed in a section called "What We're Reading." It was a semiofficial coronation from the company they religiously followed. Lenssen and his fellow aficionados have devoted fan bases and growing traffic, and many are riding the wave of the Google Economy to profitability.

CHAPTER 14

April Fools

In the spring of 2004, with business booming and Google basking in the glow of its ever-growing popularity, Larry and Sergey prepared to dazzle Internet users with a different kind of email. Building on the strong Google brand name, they called the new service "Gmail." It seemed a no-brainer to give the Google email service a catchy name that associated it with the positive vibes surrounding the search engine. Google investor Michael Moritz had been saying for several years that the two things people did most online were communicate and hunt for information. With Google dominating search, email seemed a logical next step, a great way to serve computer users and expand the scope of the business too. To keep the element of surprise alive, Google's top brass kept the details of Gmail's development a secret inside the Googleplex.

Larry and Sergey wanted to make a big splash with Gmail. There was no reason to provide the service unless it was radically better than email services already offered by Microsoft, Yahoo, AOL, and others. They built Gmail to be smarter, easier, cheaper, and superior. Otherwise Google users wouldn't be impressed, and its creators wouldn't be living up to their own high standards. This was also a chance for them to please themselves by inventing the kind of email they longed for after growing tired of using other services with all their flaws. They had identified email problems that Google, with its immense computing power, could address. For example, it was difficult, if not impossible, to find and retrieve old emails when users needed them. America Online automatically deleted emails after 30 days to hold down systems costs. There was no easy way to store the mountain of emails that an active Internet user amassed without slowing personal computers, or paying Microsoft, Yahoo, or another firm to provide additional storage.

To blow the competition away and add a Google "wow" factor, Larry and Sergey, and the Gmail team inside the Googleplex, addressed all these issues and then some. To make the new service an instant hit, they planned to give away one free gigabyte of storage (1,000 megabtyes) on Google's own computer network with each Gmail account. That was 500 times greater than the free storage offered by Microsoft and 250 times the free storage offered by Yahoo. This was a service it could afford to provide because of the way it built its own high-powered computer network. Googleware—that blend of computing power and software that had taken search to new heights—could now perform the same feat for email. One gigabyte was such an amazing amount of storage that Google told Gmail users they would never have to delete another email.

Finally, to inject Gmail with that Googley sense of magic, computer users would be able to find emails instantly, without ever having to think about sorting or storing them. A Gmail search would be fast, accurate, and as easy to perform as a Google search, making the service an instant hit among trusted employees who sampled it inside the Googleplex. They refrained from saying anything about it to outsiders, but the excitement over Gmail's public debut was mounting.

To generate the kind of word-of-mouth buzz that had made Google the leading search engine, the company decided to first give the product to about 1,000 opinion leaders to test, and then allow each of them to give away a limited number of Gmail accounts to family and friends "by invitation only." This would allow the company time to get any bugs out of the new offering. And, since it was giving away one free gigabyte of storage with each Gmail account, it would also provide a buffer to ensure that Google always remained well ahead of any demands on its computer network.

Unlike most of its new products, Gmail was designed to make money even during the test phase. With demand for advertising increasing, the company needed to increase the amount of available space it could sell. It made sense to Larry and Sergey to profit from Gmail by putting the same type of small ads on the right-hand side of Gmails that Google put on the right-hand side of search results. The ads would be "contextually relevant," triggered by words contained in the emails. It was a proven business model that served advertisers and users well as part of Google's search results. By giving advertisers more space on the Google network, Gmail would provide a healthy new stream of profits for the company that would grow over time as the communications technology caught on.

Looking at the world through Google-colored lenses, this seemed like a superb idea in every respect. It didn't occur to Larry, Sergey, or any of the other engineers in senior roles at Google that serious people they respected would strenuously object to the privacy implications of having Google's computers reading emails and then placing ads in them based on the content of those messages. In their virtual reality, they remained oblivious to the political reality that awaited them. And because they were software developers operating in something of a vacuum, they failed to get advice from the right people, or to brief opinion leaders in advance, or to do anything that would have headed off the uproar over privacy issues that Gmail sparked. At risk was nothing less than Google's good name and reputation. For the two founders, who prided themselves on being well-informed, it would be a harsh lesson in the dangers of being too clever and too cloistered.

To give Gmail an added dimension of fun and free publicity, Larry and Sergey chose to announce the hot new offering on April 1,

2004. In prior years, Google had made fanciful April Fools announcements as a joke, so when journalists and others heard about Gmail's gigabyte of free storage on April Fools' Day 2004, they wouldn't know whether it was real or made-up. That would generate more chatter and questions, heightening interest without costing the company any money on advertising or marketing. And that was just the way the Google Guys wanted it.

On April 1, 2004, Google issued a press release headlined: "Search is Number Two Online Activity—Email is Number One; 'Heck, Yeah,' Say Google Founders." The release said that the inspiration for Gmail came from a Google user complaining about the poor quality of existing email services. "She kvetched about spending all her time filing messages or trying to find them," said Larry Page. "And when she's not doing that, she has to delete email like crazy to stay under the obligatory four-megabyte limit. So she asked, 'Can't you people fix this?'"

Millions of M&Ms later, Gmail was born. "If a Google user has a problem with email, well, so do we," Brin said. "And while developing Gmail was a bit more complicated than we anticipated, we're pleased to be able to offer it to the user who asked for it." Page and Brin noted that Google would make the service available first to a small number of test users and that "with luck, Gmail will prove popular."

The press release made no mention of the ads Google intended to place in the emails, but in interviews that day, Wayne Rosing, then Google's vice president for engineering, revealed the plans. "Gmail grew out of experiments that were done that involved our ad targeting," he said. "We did some textual analysis and were able to make it work." While starting with test users and intentionally keeping Gmail scarce enough to increase its popular appeal, Rosing added, "We certainly think millions to tens of millions of users in a reasonably short period of time is a possibility."

As word spread of Google's plans to put ads in emails, politicians and privacy groups attacked the company and its plans, kicking off a media firestorm. In Massachusetts, anti-Gmail legislation was introduced. Shocked privacy advocates urged the company to pull the product immediately and began circulating anti-Google petitions. One California lawmaker threatened the company, saying that if Google didn't dump Gmail, she would press for legislation banning it. Her bill passed the state's Senate Judiciary Committee with only one opposing vote. She decried the ad-driven profiteering in emails as a gross, unwarranted invasion of privacy. For the first time, Google was being viewed with suspicion in a major way. People considered their emails private, and the notion of Google's putting ads in them based on their content seemed to cross the line.

Larry and Sergey had been blindsided. They never foresaw the prospect of hostile or negative reactions to their terrific new offering. Google's carefully honed image as a force for freedom ran headlong into visions of the search giant playing Big Brother and looking over the shoulders of millions of computer users sending emails. It was an ugly metamorphosis. At stake, commentators said, were cherished individual rights to liberty and privacy.

"Google is risking its reputation for honesty, and putting the user first, with a new email service it is currently developing called Gmail," wrote Walt Mossberg, *The Wall Street Journal*'s respected technology columnist. "The problem here isn't confusion between ads and editorial content. It's that Google is scanning your private email to locate the keywords that generate the ads. This seems like an invasion of privacy." Mossberg noted that the company said the scanning would be done by computers, but he concluded that "the proposed system is still a little creepy, and it has the potential for big problems if the content scanning were ever misused by Google. Google might also be forced to use such content scanning in the service of government subpoenas or court orders that might apply to years' worth of its customers' emails."

In a move that reflected the spreading concern, Mossberg urged Google to put out the fire. "I'm calling on Google to preserve its sterling reputation for honesty and customer focus by offering an alternative form of the new Gmail service. The company should offer Gmail accounts without the ads, and without the scanning, for a modest annual fee. That would put the choice where Google has always placed it: in the hands of its users."

Since Mossberg was a big admirer and fan of Google and its services, his column surprised Larry and Sergey. Other privacy advocates went after Gmail and Google more vociferously than the measured Mossberg. "Google is in a de facto manner creating incredibly detailed dossiers on every one of us that the government would never be allowed to create. And by creating them, they become available to the government," said Kevin Bankston, a lawyer with the Electronic Frontier Foundation, or EFF, a highly regarded civil liberties watchdog. Miguel Heft, an editorial writer for the *San Jose Mercury News*, wrote that Larry and Sergey's poor handling of the introduction of Gmail was "a bad April Fool's joke." The offer of virtually unlimited storage on Google's computers heightened the risk of governmental abuse, since emails housed on Google's machines lacked the legal protections of those stored on an individual's personal computer. Heft suggested Google redeem itself by working for the passage of strong new privacy legislation in Washington.

The privacy issue mobilized the troops. The Privacy Rights Clearinghouse, based in San Diego, and 30 other privacy and civil liberties groups from around the world signed a public letter to Google demanding that the company suspend Gmail. In addition to the U.S.-based groups, the signers included ones in Australia, Canada, Spain, the Netherlands, and Great Britain.

In their letter to Larry and Sergey, the groups urged Google to be more explicit about its policies and practices concerning the sharing of data between its search and Gmail operations. Google already retained data showing the computer address where every search originated and the nature of the query. Mixing that search data with identities and names would now be possible, since Gmail required individuals to register for the service by name. A Gmail user conducting Google searches would be potentially identifiable. All that personal information in a single electronic repository was disturbing, since it offered one-stop shopping for dishonest employees, hackers, divorce lawyers, private investigators, and overzealous prosecutors. Most people don't worry about privacy issues until their own privacy is violated. That was why privacy coalitions and groups had to speak up before it was too late.

"The Gmail system sets potentially dangerous precedents and establishes reduced expectations of privacy in email communications," said the April 6, 2004, letter. "These precedents may be adopted by other companies and governments and may persist long after Google is gone." It went on to state that it was possible that no policy Google could adopt would adequately protect consumers from future abuses, and it challenged Google's view that computer scanning of emails to insert ads was more benign than people reading private emails. "A computer system, with its greater storage, memory and associative ability than a human's, could be just as invasive as a human listening to the communications, if not more so."

Adding to fears was Google's offer of free hefty storage and permanent email retention. This meant that emails would be retained indefinitely on Google servers, even though the strongest privacy protections in the U.S. would be exhausted after 180 days. For these and other reasons, the Electronic Privacy Information Center issued a statement concluding that Gmail was a defective product that violated the "sanctity" of private communications. Google intended to retain and store records of emails and searches, creating a tempting database of personal information about millions of people around the world. For many, part of the beauty of Google searches had been their apparent anonymity, but revelations about the company's retention of logs of traceable search requests called this into question.

Amid the furor over privacy, the free Gmail accounts, in intentionally short supply to make them more appealing, began trading for as much as $100 each on the eBay auction site. When Larry and Sergey heard that news, they sensed the backlash would eventually blow over. After all, Google as an entity was about pushing the boundaries and limits of innovation, and sometimes it took people time to adjust to change. In any event, privacy groups frequently squawked about things that didn't bother most people, who favored convenience, performance, and price. And besides, as they repeated to themselves inside the Googleplex, people must know that the other major email services have computers that scan the electronic messages for pornography, viruses, spam, and more. Looked at that way, Gmail was no different. Those were the facts.

Because from their perspective this was much ado about nothing, Larry and Sergey saw no need to be defensive or respond to crazed critics. In fact, all the publicity would certainly heighten awareness of the Google search engine and its Gmail progeny. Soon enough friendly columnists who tested Gmail and fell in love with it would begin writing about why the outcry was unjustified. Tradition-bound companies might have seriously considered pulling Gmail, at least temporarily, to quell the uprising. But this was Google, and it had the clout, and confident leadership, to ride this out without flinching. The founders began to respond on-message.

"It sounded alarming but it isn't," Sergey said. "The ads correlate to the message you're reading at the time. We're not keeping your mail and mining it or anything like that. And no information whatsoever goes out. We need to be protective of the mail and of people's privacy. Any Web mail service will scan your email. It scans it in order to show it to you; it scans it for spam. All we're doing is showing ads. It's automated. No one is looking, so I don't think it's a privacy issue. I've used Gmail for a while, and I like having the ads. Our ads aren't distracting. They're helpful."

When Google tested Gmail, people bought lots of things by clicking on the ads. To Larry, this was proof that computer users, advertisers, and Google's coffers were all well served by the small ads on the right-hand side of a Gmail. "Even if it seems a little spooky at first, it's useful," he said.

But as the anti-Gmail forces gained strength, not everyone at Google was sanguine about the uproar. Some of the company's major investors were furious about the way Gmail was rolled out—and about the timing. How could CEO Eric Schmidt have allowed this to happen just weeks before the private company was set to reveal its plans to go public by selling stock? He was supposed to prevent the Google Guys from doing things they regarded as really smart but the rest of the mere mortals on the planet would consider dumb, or at least ill-timed. The Gmail fight had the potential to give Google a black eye and damage its credibility. Left unchecked, Gmail could undermine the single most important asset that Google had: the trust of hundreds of millions of computer users and advertisers around the world. It had the potential to destroy years of goodwill. Moreover, while Google had brought this upon itself, Microsoft, Yahoo, and other competitors would seize the opportunity to keep the negative spate of stories alive. They would put the word out that Google really was different. For example, it retained copies of all emails, even those that individuals deleted.

Page's view was that Google's biggest blunder was not the ad features in Gmail, but the way it had rolled out the new email service. "We learned a few things," Larry said. "We could have done a better job on the messaging. People started talking about it before they could try it. I didn't expect them to be so interested. We released the privacy policy, and they were very interested in that. It was all they had access to, so it sparked a lot of controversy."

Either Larry or Sergey needed to get more involved if they were to stop the anti-Gmail forces, and to do that, they needed a bona fide privacy group who could advise them—one that would tell them the truth, no matter how dire, without being alarmist. Although he wanted to focus on users, Larry recognized the need to view Gmail from a political and privacy perspective. He had a friend in the privacy world, Brad Templeton, and it was time to reach out to him. Templeton was a man who had been around cyberspace so long, and had been at the birth of so many breakthroughs, that he was literally credited with putting the dot in the ".com" when initial decisions were being made about addresses for the Internet. Templeton wore a lot of hats: he was chairman of the Electronic Frontier Foundation, a leading privacy advocate; he was Larry's friend and Burning Man compatriot; he had consulted for Google; and he earned money from having Google-delivered ads on his personal Web site. It was time to turn to him for counsel.

As Page saw it, if there was an honest broker in the Gmail debate, it was Brad Templeton and the Electronic Frontier Foundation. The group Templeton chaired had declined to join the coalition calling on Google to scrap Gmail, though Templeton viewed the immediate privacy issues and their long-term implications very seriously. Given his liberal upbringing, Page understood the importance of striking a balance on issues like these. Templeton, for his part, had come to think of Gmail as complex, not as good or evil. It fell into a gray zone, depending on various circumstances. He discussed the matter with his colleague Kevin Bankston at the EFF. Bankston felt it made sense for people worried about privacy to take matters into their own hands by using one firm for email and another company for search; that way, the information about them would not all be in one place. "There are dangers of one-stop shopping for online services from a privacy perspective," he said. "The best thing to do is to break it apart. I would not use my primary email provider for search."

Templeton met with Page to review the implications of Gmail, and then developed a thoughtful, even-handed analysis of its pros and cons. "[Google] should encrypt email so it is technically impossible to search the entire archive," he said. "They should distill and strip personal information from their logs. The steps Google can take involve sacrifices so they haven't taken them. They are doing research and like to go and look back at patterns of users and customers. So they are not destroying it."

Templeton addressed the Google privacy clash artfully. "While there has been over-reaction to Gmail," he wrote, "there are some real issues here to be worried about." These included, under current federal law, the loss of legal protections for email retained by Google for more than 180 days. He said that correlation of Gmail and Google searches—something Google was doing in order to customize and personalize services for computer users—poses enormous privacy risks, and these would only become more pronounced as more people used Gmail and more of their lives moved online. "In the modern era where computers threaten privacy, we are as afraid of outside computers knowing all about our lives as much as we are outside people." One reason for this, Templeton stated, is the reality that information stored on external computers has sometimes found its way out into the world and fallen into the wrong hands. By having emails and searches stored on Google's computers outside of their homes, people were inadvertently building the database for what he termed a "surveillance society."

"When our papers are at home, mass surveillance of them simply doesn't scale. It's too expensive. They must be broken into a million times. Online, it scales well. Those outside machines must be compromised (by the law or system crackers) only once."

Templeton also cited global issues that Google had not addressed in considering Gmail in advance of its release: the way that governments in different countries would respect, or invade, the Google database based on their own cultures and laws. "When we build our systems," Templeton said, "how often do we think of what it will mean when they become popular in China or Saudi Arabia? Or what it means when they are sold to other companies whose policies are not so benevolent?"

Most of all, Templeton said, computer users themselves needed to be aware of the perils they could face. "You should have no more expectation of privacy in an email than you have in a postcard, or worse, a postcard you hand to a third party to carry." Many of his own Google searches involved private information, he said, including looking up the names of prescription drugs he has been given. The biggest danger, he concluded, was the merging of Gmail and Google searches in a climate in which civil liberties were being stripped. The illusion of privacy was being replaced by the appearance of surveillance. The problem with this, Templeton observed, was that it had a corrosive effect on people's personal freedom and their willingness to express themselves. For these reasons and more, the mere fear of Google and Gmail was an issue in itself. "It is not only important to have your privacy. It is important that you believe you have your privacy. If you even suspect that you are being watched, it changes your behavior and you become less free as an individual."

But Templeton also noted that with credit agency records, bank records, medical records, and more stored on corporate computers already, it was inevitable that additional personal information would move online. And while it wouldn't be perfectly protected, people would adapt. Meanwhile, a number of other privacy advocates who had attacked Gmail tried it and began using the new service avidly. Journalists praised it too; for the first time, they could find old emails they were looking for easily and quickly. The storage was free and immense. And Gmail had an interesting way of turning a series of emails between two people into something resembling a conversation. Larry and Sergey anticipated that as more people actually experienced Gmail, its strengths would win them over and privacy concerns would fade. "Gmail is a very nice product, with great promise," Templeton said. "Much of what Google is doing with Gmail is innovative and worthwhile."

CHAPTER 15

Porn Cookie Guy

Matt Cutts, who used to work at the National Security Agency on encryption and security issues, now fights unwanted pornography for Google instead. Cutts, whose nickname is "Porn Cookie Guy," developed and maintains Google's SafeSearch filter, designed to block pornography for computer users who activate it. Cutts got his moniker by giving out his wife's tempting homemade cookies to Googlers who help him find unwanted porn. A Kentucky native and University of North Carolina Ph.D. candidate, Cutts is charged with finding and eliminating unwanted porn both in Google search results and in ads. Marissa Mayer, Google's director of consumer Web products, said that what was once a giant problem for Google has been reduced dramatically.

"Matt does a lot around our search quality to make sure that various surprises don't happen to you," Mayer said. "For those of you who don't want offensive results, or potentially offensive results, to appear in your query, he has worked to remove those. Even for people who don't have SafeSearch turned on, he has removed spurious porn."

Still, sneaky porn sites find ways to show up in Google. Their owners often purchase the expiring names of other Web sites as a way to get displayed in Google search results without revealing the real nature of their content. "There are people who buy hundreds of domains," Cutts said, a practice known as "porn-napping."

While combating unwanted porn, Google makes millions of dollars annually on pornography ads displayed alongside search results. One out of every four requests for information on Google and other Internet search engines involves pornography, according to a 2004 study by Family Safe Media, a watchdog group. That statistic suggests that Google fields tens of millions of requests for pornography daily.

Google Image Search, the tab above the search box on the homepage, displays graphic sexual content for free. On the right-hand side of the results pages is a string of paid ads that appear every time a computer user enters a search term for adult content. "Google generally allows ads containing adult themes, such as explicit sexual content," the company states on its Web site. Because of these policies and features, and its position as the largest and most popular search engine, Google is one of the largest single gateways on the Internet for pornography. The company does not disclose how much money it makes from pornography or any other single business category. But the anonymity and the relative ease of accessing pornography has obvious appeal to some adults and underage adolescents, since they can avoid engaging in conspicuous consumption, the 2004 study said.

Computer users searching for pornography online may be mistaken in concluding that they are viewing it anonymously and privately. Google maintains electronic records of all searches, which can be traced back to specific computers. If someone has a Gmail account or has registered for any other Google service, the firm's electronic records could be used to trace porn searches to specific individuals.

Not surprisingly, both Google and its biggest competitor, Yahoo, profit handsomely by selling sex-related ads. Pornography, within limits, is protected by the First Amendment in the U.S. In countries where online pornography is banned, including Germany and India, Google and Yahoo abide by local statutes.

However, not all of Google's ad policies are vigorously enforced. For instance, Google has stated that the search engine will not accept ads for prostitution, yet it sells ads to Web sites that bid for the phrase *escort service*. Typing those words into Google yielded the following ad: "Hot Casual Sex. Men & Women in Your Area. Find the Match You are looking for now." Typing "XXX" into Google yielded: "Millions of People looking for Sex! Sign up today and get laid tonight." On Google's Web site, it states that SafeSearch blocks Web pages containing explicit sexual content from appearing in search results. Using the SafeSearch filter at its strictest level, a search for "XXX" blocked all adult content from appearing in free search results, but it did not block ads with adult content, including one that read, "Free Webcam Community: Live Hot Girls, Free Video Chat," and another with the headline "Sexy Girls and Sexy Guys."

Typing the word *pornography* into Google produces free results featuring antipornography Web sites, while typing *porn* leads to adult content. In both cases, to the right of these search results are small text ads that Google has sold to promoters of pornography on the Internet, who pay it a fee every time somebody clicks on one of their "hot, hot, hot" ads. At the same time, Google seeks to carefully abide by all federal and state laws restricting child pornography. Advertising on Google is not allowed for the promotion of child pornography or other nonconsensual sex, the company says.

Google's sale of adult-content ads is part of a larger set of idiosyncratic standards. These choices offer a window into the Don't Be Evil philosophy and values of Google's founders, especially Sergey Brin, who has the power to decide what kinds of ads the search engine will accept or reject. The ad policies generally reflect the personal preferences of Google's founders, and the firm publishes a detailed policy statement on its Web site about ads.

Google accepts ads for wine, while rejecting ads for beer and hard liquor. As a matter of policy but not always practice, it rejects political attack ads that disparage an individual running for office, but accepts hostile ads that distort and demonize the positions of an elected official or candidate. It rejects all cigarette advertising, and makes a further political statement by flatly refusing to accept gun ads. But it does profit from the sale of advertising for certain bullets, silencers, and other devices that can make guns even more lethal. This is despite a written policy which states that it does not accept ads for ammunition.

Google further states that it does not accept ads promoting drug cleansing shakes, urine test additives, or other products designed to help people pass drug tests. It does not permit ads that promote violence or advocate against a "protected group" as distinguished by age, race, religion, disability, or sexual orientation. It rejects ads for "black boxes" and other devices that enable people to get free cable television. Google's policy also prohibits advertising to promote illegal drugs and drug paraphernalia, fireworks, online casinos, miracle cures, radar detectors, and various weapons, including brass knuckles.

In early 2004, Google took legal action against Booble.com, a Web site designed as a search engine for sexual content. Google said the similarity in the names, look, logo, and colors on Booble and Google could lead to consumer confusion and violate trademark restrictions. Booble, a British site that bills itself as an adult search engine and pornography directory, fought back, saying it had the right to do a Google parody under the First Amendment.

Booble's logo replaces the *o*'s in the middle of its name with a pair of breasts. And instead of the "I'm Feeling Lucky" button found on Google, Booble offers users a button that says "I'm Feeling Playful."

CHAPTER 16
Going Public

Larry and Sergey had put off taking Google public for as long as they could, but a late April 2004 deadline was fast approaching. The advantages of remaining a private company were enormous, and they hated giving them up. Worst of all, rivals Microsoft and Yahoo would learn just how profitable Google had become, as well as many details about the scope of the company's operations. Once that information became public, competition would heat up. But federal rules required public disclosure of financial results by companies that had a substantial amount of assets and shareholders, and Google had exceeded those limits. The company had also recruited new hires by giving them plenty of stock in the private firm, and Larry and Sergey felt an obligation to the employees to provide them with a way to convert their holdings to cash.

They knew the corporate landscape was littered with the debris of companies whose founders and employees had lost their way after working hard and smart while the company was private. In many cases, the unfortunate result of going public was that intensity and focus waned as hundreds of employees, many of whom could not afford to buy their own cars with cash before, suddenly became millionaires. Then, too, there would be immense unwanted publicity surrounding Brin and Page. Everyone would know they were billionaires, and they worried about their lifestyles and the safety of their families. What would happen to the freedoms they enjoyed in their day-to-day lives? Would they need bodyguards watching their every move to protect them from harm? Sergey's dad had continued teaching mathematics at the University of Maryland, and his mother had continued working at the Goddard Space Flight Center. Many of their fellow professors and rocket scientists had no idea about the Google connection. Would the knowledge of Sergey's wealth change life for his parents or endanger them in any way?

As much as they hated to admit it, Brin and Page knew down deep that they had taken the first step toward an inevitable public offering on the day they accepted $25 million from Kleiner Perkins and Sequoia Capital. The venture capital firms had a duty to realize a return on stakes they purchased in companies, and to return money or stock to investors in their funds. At the time, the founders had needed the money to ramp up Google's growth rate, just as they had needed that initial $100,000 from Andy Bechtolsheim to buy the batch of computers and parts that began the true build-out of their firm. There was also the $1 million they had raised from family, friends, and the cadre of angel investors who could cash in through an IPO.

But while it would be nice to have additional cash in their corporate war chest to use in the growth of the business and for the anticipated clashes with Microsoft, the company was already generating plenty of money. And neither Brin nor Page really needed the billions of dollars they would personally pocket from taking Google public, since both lived relatively modestly and simply. Nor did they care much about the accumulation of wealth as a scorecard or means of measuring their success.

For most Silicon Valley entrepreneurs, an IPO was the ultimate dream, a time to bask in the limelight and measure their worth the American way: with dollar signs. But Brin and Page were just the opposite. They loved the privacy, they loved the freedom, and they relished having analysts and competitors consistently underestimate Google's performance. Since the company was debt-free, self-funded, and had plenty of cash, they didn't need to sell stock to the public to raise money. The only upside they saw was that they would have more resources to grow and realize their vision for Google.

If they were going to go public and give up their cherished privacy, Larry and Sergey agreed to do it their way, just as they had when raising their venture capital. Neither Wall Street nor anybody else was going to tell them how to do the deal. It didn't matter that high finance was outside the realm of their core expertise. The Wall Street stuff seemed simple compared to building a great search engine, motivating employees, and running a rapidly growing, profitable company, so as far as they were concerned, they would find a way to maintain total control over Google and over the IPO too. This mattered more to them than filling up their bank accounts. They might have no choice about doing an IPO, but that didn't mean they were going to go running to Wall Street hat in hand to ask for help.

In the annals of Wall Street, no business had ever done a successful billion-dollar IPO the way Larry and Sergey wanted to do it. That didn't scare them at all. Accustomed to dreaming and doing things nobody else dared, they were determined to blaze a new trail with their IPO. They were going to ask questions, get answers, and make decisions based on what they believed was right. If anybody on Wall Street didn't like it, they could sit this one out.

On the eve of a traditional IPO, a Wall Street investment firm, based on investor demand, market conditions, and other factors, sets the initial price for selling a company's stock to the public. The tension between the issuing company and the investment firm is clear. The Wall Street brokerage house has an incentive to underprice the offering so that it would be easier to sell and would enrich favored investors when the stock price shot up on the first day of trading. The issuing company, on the other hand, wants the highest possible price, because it is raising money for its business through the sale of shares. The higher the price, the more money a business raises. Wall Street investment bankers sought to persuade corporate executives that they were better off to "leave some money on the table" by pricing the IPO a bit low. That way, they argued, big investors would feel good about the deal and be more willing to buy shares later, if and when the company needed to raise additional money.

The major Wall Street firms controlled the entire IPO process. They set the initial price of the stock, decided which investors would be allocated shares, and charged princely fees for their services. Larry and Sergey knew about the scandals—some criminal, some unethical, all evil—where Wall Street underpriced IPOs and then let favored clients profit by dumping the stock on day one, after the price skyrocketed. The guys wanted no part of what seemed to be a corrupt and rotten system.

Supremely confident, they placed their trust more in mathematical equations, software, and technology than they did in Wall Street advisors hungry for fees. In fact, in the Internet age, they had a hard time understanding why Wall Street was still raising money for companies the old-fashioned way, using the same basic process it had used for decades. It was as if technology had come along that could do the job better, but the Wall Street crowd preferred to do business the way it always had—putting a premium on relationships and fees rather than on what was best for each client. As long as no major investment house changed its ways, none of the other firms would be forced to change either.

From Wall Street's vantage point, a company like Google that was powerful, brash, and well-known enough to attempt to set its own rules for going public was rare. More often, businesses needed road shows, hand-holding, advice, and a proven way to get investors to listen to their story in order to get financings done. For these relationships and services, companies were willing to pay fees that, Wall Street advisors told them, would come from the money raised in the offering, rather than directly out of the company's coffers.

However, in its financial offering statement filed with the SEC, Google outlined an entirely different method for distributing stock to the public, which it viewed as egalitarian, meaning anyone could participate. That would overcome the Wall Street bias toward underpricing. The process involved a high-tech version of something called a "Dutch auction," which drew its name from the way growers priced and sold tulips and other flowers in Holland.

Just as Google sold ads based on a continuous automated auction, the company would price and sell its stock based on bids received online from potential investors. Everyone who came in at a level at or above the so-called clearing price set by the company would be allocated stock. Everyone below that price would not. (Google executives later said they would have preferred to do a continuous stock auction with no fixed price, but SEC rules required setting one IPO price for all investors.) Google, before receiving bids electronically from investors who placed orders with brokerage firms, would publicly announce a minimum and maximum price and invite people to bid with that range in mind. Major investors, and small ones too, would be eligible to buy shares through the same process. There would be no favoritism, no family-and-friends allocations of shares, and no sweetheart deals. Instead, even novices—with little money to put up and who normally were ignored by Wall Street—could participate in the auction, provided they could afford to buy at least five shares. That was an unusually tiny minimum, which brokerage houses would not have allowed if Google had not insisted on it.

Suddenly, it appeared that millions of Google users in the U.S. who had never participated in an initial public offering before could buy a handful of shares, if they could afford them, and not be shut out just because of where they lived or whom they knew. It was quintessentially Google. Brin and Page, reluctant as they were to take the company public, decided they would do so in as populist a fashion as they could.

Brin and Page resented what they viewed as a Wall Street monopoly when it came to fees. All the firms charged the same exorbitant fees for handling IPOs, whether the deals were easy to sell, hard to sell, or routine. The firms they interviewed all spoke of a 7 percent fee; in a $2 billion public offering, they would earn $140 million. In theory, they pocketed these fees in exchange for the risk that they would be left holding the stock. This was the concept behind "underwriting": Wall Street firms guaranteed companies doing IPOs that they would receive a specified price for their shares, and were in a sense being paid for the risk that they would be unable to unload the stock for one reason or another. But in practice, all the firms charged the same fees on every deal and presold large chunks of shares. This made no logical sense to Larry and Sergey, especially not for a desirable stock like Google's.

The founders decided they would compensate Wall Street at less than half of the usual and prevailing fees, and if the brokerage houses didn't like it, they didn't have to participate in the deal. Furthermore, the guys developed detailed plans to wrest control over the unfair pricing and allocation of shares that had created scandals on the Street, and reserved the right to cancel the deal at the last moment if they changed their minds. To put it mildly, the Google Guys were sending a "drop-dead" message to Wall Street. If the company succeeded, the deal could potentially slash the fees and the role of the middleman in other initial public offerings.

Larry and Sergey would also decline to name anybody to serve as chairman of Google's board of directors in the process of going public. The post would simply remain vacant. This was highly unusual, but it was another way for Brin and Page to retain control. CEO Eric Schmidt would serve as chairman of the board's executive committee, enabling him to carry out the necessary ceremonial and legal duties that came with being a public company. Brin and Page would remain co-presidents and controlling shareholders. And the two of them, with Schmidt on a short leash, would operate Google together. They would name a board chairman later, if and when they got around to it.

Every Wall Street firm that met with Google to discuss the possibility of handling its IPO had to sign a serious confidentiality agreement. After Credit Suisse First Boston and Morgan Stanley were chosen to comanage the stock offering, Google demanded that they sign new confidentiality agreements at every meeting. On top of that, the company revealed as little financial and operational information as possible to the firms, keeping them in the dark for as long as possible. And Google also put all of Wall Street on notice that the legal consequences of any leaks about anything, before or after the IPO, would be severe, prompting investment bankers and lawyers to grumble that they had never seen anything like these impossible Google people before.

Among other things, Larry and Sergey were advised by their outside lawyers at Wilson Sonsini Goodrich & Rosati, a prestigious firm that had handled virtually all of the biggest transactions involving Silicon Valley and Wall Street, that once they filed IPO documents with the Securities and Exchange Commission, they would enter something known as the "quiet period," when they would have to make sure they said nothing to tout the value of Google stock. As populists, it made no sense to Brin and Page that the quiet period permitted something called a road show, where they would meet behind closed doors with the big money people, the institutional investors and heavyweights of Wall Street, to give presentations and answer questions. What about the average investor? What about the typical Google user who might like to invest? Why did it make sense to give those big insiders an advantage at road show gatherings from coast to coast, leaving outsiders and small investors to fend for themselves? This seemed like a typical self-serving Wall Street tradition, and they aimed to break it, or at least to bend it, in the process of going public. They would say little or nothing new at the road shows, and give everyone access to the same additional data about Google by posting it on the Internet.

Google is not a conventional company. We do not intend to become one.

Thus began a letter from Brin and Page accompanying the disclosure of financial and operational details about Google in its mandatory IPO filing with the Securities and Exchange Commission in the third week of April 2004. The idea of including a philosophical letter from the founders with the required SEC disclosures had been approved by Wilson Sonsini. The founders' letter was instantly made available electronically to anyone with an Internet connection. It was the sort of thing the pair was determined to do, no matter what their venture capital backers, John Doerr and Michael Moritz, really thought about it. The letter had been Larry and Sergey's idea from the start. They wanted to show that Google had personality, that it was a different type of enterprise and place to work. Most other public companies filed mundane IPO documents filled with standard legal and financial disclosures. The Google Guys wanted to snare the world's attention with an unusual letter about the company's culture and their view of the world.

Worried about the potential consequences of the draft letter warning Wall Street and any investor who disagreed with Brin and Page to stay away, Moritz went to work. The night before its release, he finally wrested a copy of the letter away from Page. With a brushstroke here and a keystroke there, he gave the letter a more measured tone, blunting some of the sharpest edges, adding semantic nuances, and, critically, incorporating more fully the role that CEO Eric Schmidt would play in the company's management.

Having forced the Google Guys to hire Schmidt, the last thing Doerr and Moritz needed was to have his stature diminished in the eyes of investors. To have confidence in the company's future and achieve the right valuation in the IPO, Doerr and Mortiz knew from experience that investors needed to be convinced that while the founders drove innovation and sought to change the world, a steady, experienced executive would be there to ensure that the public company focused on stockholders, put in a system of checks and balances, and operated in a reasonable, financially sound manner.

Over and over, Brin and Page stated in the letter that they fully intended to do the same things as a public company that they had done successfully while Google was private. For instance, they would not kneel at the altar of Wall Street's holy grail of quarterly profits. Instead, they would do whatever they thought was best for Google in the long run.

A management team distracted by a series of short term targets is as pointless as a dieter stepping on a scale every half hour, they wrote. *In Warren Buffett's words, "We won't smooth" quarterly or annual results: If earnings figures are lumpy when they reach headquarters, they will be lumpy when they reach you.*

They called their letter "An Owners Manual for Google Shareholders," and said it was inspired by the letters accompanying the annual reports of Berkshire Hathaway, the giant insurance company headed by investment guru Warren Buffett. With a stroke of the pen, Larry Page and Sergey Brin had associated themselves with Buffett, the most successful American investor of the era.

Google was a pure-play technology firm that, unlike some rivals, including Yahoo, did not own or generate original content. But it earned its profits from advertising, a classic trait of a media company. In their letter the founders laid out plans for Google to issue two classes of stock: Class A shares for regular investors, which carried one vote each; and Class B stock for themselves, carrying ten votes per share and giving them absolute control. The dual class structure would make a takeover of the company impossible without their approval, would discourage public investors from seeking to influence management, and would facilitate their ability to run the company without interference. This was the ideal way for them to stay in charge while Google raised billions of dollars by going public.

Brin and Page justified the two classes of stock with unequal voting rights by comparing six-year-old Google to three of the leading newspapers in the U.S.—*The Washington Post*, *The New York Times*, and *The Wall Street Journal*—and the multiple classes of stock used by the families that controlled these newspapers in order to preserve editorial independence. For a pair of iconoclastic thinkers who loved Silicon Valley and rejected the traditional ways of Wall Street, when it came down to making the most convincing and persuasive case possible, it was a matter of not confusing principle with preference.

The main effect of this structure is likely to leave our team, especially Sergey and me, with increasingly significant control over the company's decisions and fate, as Google's shares change hands. The New York Times Company, The Washington Post Company and Dow Jones, the publisher of The Wall Street Journal, *all have similar dual class ownership structures. We believe a dual class structure will enable Google, as a public company, to retain many of the positive aspects of being private.*

The proposed dual class structure lacked one of the essential ingredients that Brin and Page sprinkled inside Google to maintain the work ethic of their employees: accountability. To put it simply, virtually everyone hired to work there had to be approved by Brin or Page or someone senior, and they even insisted on seeing transcripts and test results. The founders, who had the ultimate power to hire and fire, had faced no similar system of accountability when they ran Google as a private company, and they saw no need to change that as the firm made the transition to public ownership. This meant they could dismiss Eric Schmidt as easily as the most recently hired software engineer.

As an investor, you are placing a potentially risky long term bet on the team, they wrote. *We believe a well functioning society should have abundant, free and unbiased access to high quality information. Google therefore has a responsibility to the world. The dual class structure helps ensure that this responsibility is met.*

Brin and Page said they ran Google with a motto in mind: Don't Be Evil. And as far as search was concerned, they explained just what they meant.

Our search results are the best we know how to produce. We do not accept payment for them or for inclusion or more frequent updating. We also display advertising, which we work hard to make relevant, and we label it clearly. This is similar to a well-run news-

paper, where the advertisements are clear and the articles are not influenced by the advertisers' payments.

With this immodest declaration, they were staking out their turf and taking shots at Yahoo and Microsoft, their main competitors. The Google Guys were labeling Yahoo—the popular Web site and number two search engine in the U.S.—as "evil" for accepting payments from Web sites to improve their chances of showing up in searches. It was a mantra with a message. Google search results were good and pure; Yahoo's were tainted.

But the distinctions were not that simple. It turned out that most computer users did not realize that Google's search results even contained ads, according to a study by the Pew Charitable Trusts. This was a major reason why many bright people didn't understand how the company made money. The Pew study said that 62 percent of Google users did not understand the difference between its free search results and the ads it displayed to the right of these. If more people realized that the small square boxes of text were paid advertisements, they would be less likely to click on them, according to marketing experts. Google's profits were growing faster because of ambiguity in a new medium.

By labeling the ads as "Sponsored Links," Google avoided being precise. This term lacked the stigma of "Advertisements," and as a result, more people clicked. "Google's ads are unusually effective because most people don't realize they are ads. Is that evil?" asked Alan Deutschman in an article in *Fast Company* magazine.

To many, Google's comments about good and evil seemed totally self-serving, an assertion that the Google way was the only way. Still, it set the company apart and garnered global attention. It also had a positive influence on potential Google hires and employees—one of whom wrote about it on a whiteboard inside the Googleplex. Many of the best engineers had a strong sense of the deeper philosophical issues of right and wrong, and of good and evil. Technology, in and of itself, could be a force of light or darkness. By instinct, talented technologists were attracted to a company that had appealing values and virtues that went beyond maximizing profits and market share. Following the multiyear antitrust proceedings and lawsuits that had painted Microsoft and Bill Gates as greedy monopolists, Google's stance carried great weight.

Google's remarkable financial performance, revealed in its IPO filing, stunned analysts, competitors, and investors. The speedy search engine had a penchant for profit. In the first half of 2004, the company recorded sales of $1.4 billion and profits of $143 million, compared to sales of $560 million and profits of $58 million in the same period in 2003. The trajectory suggested accelerating momentum. If the financial results disclosed by the company had not been so stunning, the words flowing from Brin and Page would not have been as noteworthy. But in a world where people listen more closely to the rich and mighty than to the poor and weak, the financial results gave currency to the arguments made by Google's founders. The firm had chosen to remain private in the late 1990s when so many Internet companies with so little in the way of sales, and no profits, went public. Google had waited until the last possible moment to reveal itself—and in the interim it had built a money machine that garnered headlines around the world, leaving investors salivating.

"The century's most anticipated IPO was on, and the document revealing the search giant's financial details, business strategy and risk factors instantly eclipsed Bob Woodward's Iraq book as the most talked about tome in the nation," according to *Newsweek.*

Brin and Page had the money and idealism to put the financial nitty-gritty aside. In their description of goals for Google, the entrepreneurs said they hoped its prosperity and ingenuity would be applied to solving major world problems. *We aspire to make Google an institution that makes the world a better place. We are in the process of establishing the Google Foundation [and] intend to contribute significant resources to the foundation, including employee time and approximately 1 percent of Google's equity and profits. We hope someday this institution may eclipse Google itself in terms of overall world impact.*

Google's IPO filing gave SEC officials fits, however. In a series of private letters, they peppered the company with detailed questions about the mechanics of the stock auction and also challenged the philosophical founders' letter, criticizing its folksy flavor.

"Please revise or delete the statements about providing 'a great service to the world,' 'to do things that matter,' 'greater positive impact on the world,' 'don't be evil,' and 'making the world a better place,'" the SEC wrote to Google and its lawyers. "Revise the section under 'Making the World a Better Place' to describe any negative perceptions associated with your products, such as privacy concerns related to your Gmail service." The SEC also had dozens of financial and legal questions, some based on items that might give investors a false or incomplete impression of the risks surrounding Google. "Your statements that the Overture Services lawsuit is 'without merit' is a legal conclusion, which Google is not qualified to make," the SEC wrote. "Please revise [or] omit this statement."

The commission would not bless this deal without revisions to the filing. While Google made certain concessions, Larry and Sergey were not going to abandon its touchstones either. The SEC officials particularly disliked the way that the founders, Schmidt, and others were referred to so casually. "Throughout the document, you refer to executive officers, directors and principal shareholders by their first names," they wrote. "For clarity, please consider revising the disclosure to refer to these persons by their full names or by their last names." Larry and Sergey refused.

CHAPTER 17

Playboys

The trouble started on May 4, 2004, only days after Google's celebrated coming-out party. Geico, the giant automobile insurer, filed a lawsuit against the search engine for trademark infringement. The insurer claimed that Google's advertising system unlawfully profited from trademarks that Geico owned. Since all of Google's revenue and growth was from advertising, the disclosure of the lawsuit appeared ominous. "We are, and may be in the future, subject to intellectual property rights claims, which are costly to defend, could require us to pay damages, and could limit our ability to use certain technologies," Google disclosed in a public filing outlining potential risks. Abroad, where Google had promising growth prospects, similar court challenges also arose. "A court in France held us liable for allowing advertisers to select certain trademarked terms as keywords," the company declared. "We have appealed this decision. We were also subject to two lawsuits in Germany on similar matters."

To make matters worse, it turned out that prior to its IPO filing, Google had eased its trademark policy in the U.S., allowing companies to place ads even if they were pegged to terms trademarked and owned by others. That was a significant shift, and one Google warned could increase the risk of lawsuits against the company. It was also a practice that Yahoo, its search engine rival, did not permit. Google claimed it made the policy change to serve users, but some financial analysts said it appeared designed to pump profits before the IPO.

And there was more. Competition from Yahoo and Microsoft posed a greater challenge to Google following the disclosure about its mammoth profitability. With so much money at stake, the intensity of the competition would heat up. Such competition might be good for computer users searching the Internet, but Google said it posed additional risk for potential shareholders. "If Microsoft or Yahoo are successful in providing similar or better Web search results compared to ours or leverage their platforms to make their Web search services easier to access than ours, we could experience a significant decline in user traffic," the company disclosed. In addition, Google warned that it's momentum seemed unsustainable due to competition and "the inevitable decline in growth rates as our revenues increase to a higher level."

Then there was the question of Google's exclusive reliance on advertising, and one particular type of advertising, for all of its revenue. That was potentially quite problematic. If Yahoo or Microsoft gained ground on search, users could flock to their Web sites, and advertisers could follow. "The reduction in spending by, or loss of, advertisers could seriously harm our business," the company disclosed in its SEC filing.

In the beginning, the firm earned all of its money from ads triggered by searches on Google.com. But now, most of its growth and half of its sales were coming primarily from the growing network of Web sites that displayed ads Google provided. This self-reinforcing network had a major stake in Google's successful future. It gave the search engine, operating in the manner of a television network providing ads and programming to network affiliates, a sustainable competitive advantage. But there was a dark side there too, because of the substantial revenue from a handful of Google partners, notably America Online and the search engine Ask Jeeves. If at any point they left Google and cut a deal with Microsoft or Yahoo, the lost revenue would be immense and difficult to replace. "If one or more of these key relationships is terminated or not renewed, and is not replaced with a comparable relationship, our business would be adversely affected," the company stated.

Google's small, nonintrusive text ads were a big hit. But like major television and cable networks, which were hurt by innovations that enabled users to tune out commercials, the company faced the risk that users could simply turn ads off if new technologies emerged.

Going public also posed a potentially grave risk to Google's culture. Life at the Googleplex was informal. Larry and Sergey knew many people by their first names and still signed off on many hires. With rapid growth and an initial public offering, more traditional management and systems would have to be implemented. No more off-the-shelf software to track revenue on the cheap. Now it was time for audits by major accounting firms. As Google's head count and sales increased, keeping it running without destroying its culture was CEO Eric Schmidt's biggest worry.

Google, the noun that became a verb, had built a franchise and a strong brand name with global recognition based entirely on word of mouth. Nothing like it had been done before on this scale. The Internet certainly helped. But Google's profitability would erode if the company were forced to begin spending the customary sums of money on advertising and marketing to maintain the strength of its brand awareness. Marketing guru Peter Sealey said privately that the advice he gave Google to study consumer perception of the Google brand was rejected by the company and that they were unwilling to spend money on marketing.

"They're technologically arrogant about the need to craft the brand and communicate with consumers," Sealey said. "These guys don't even know what their brand stands for. They are code writers."

Though pornography was big business on the Internet, Google for a time had claimed that it ran no adult ads. Prior to the IPO, the search engine's automated sign-up system no longer blocked all of those commercial messages, resulting in an additional legal risk. The company also faced potential liability from ads it ran for pharmaceuticals, financial services, and alcohol or firearms.

If the experience of other newly public companies was any indication, it was possible that a Google brain drain could follow its IPO. The generous way Sergey and Larry handed out stock options to recruits while the search engine was private, as well as looser-than-normal restrictions on selling, made cashing in and leaving the company easier. In Silicon Valley, new start-ups attracted such talent. Also, hundreds of Google millionaires might lose motivation and focus. Could the company ever create the sort of financial incentives necessary to recruit fresh talent, in the absence of the huge financial rewards reaped by early employees? Finally, what if Sergey and Larry decided Google wasn't a fun place anymore and left to pursue other ventures? "If we lose the services of Eric, Larry, Sergey or our senior management team, we may not be able to execute our business strategy," the company warned.

Worst of all, Google was the target of a billion-dollar lawsuit filed by Yahoo's Overture subsidiary, which charged that Google's entire advertising system blatantly infringed on the one that Overture had patented. Resolving the litigation could prove costly at best, requiring the company to pay hefty one-time or ongoing licensing fees to Yahoo. At worst, it could force Google to seek another ad model.

Meanwhile, the Securities and Exchange Commission had launched a probe into the company's internal procedures. The search engine had issued enormous quantities of stock and options without registering the shares or revealing its financial results to its private employee-shareholders. A pending investigation like this one by the SEC before a public offering could be the kiss of death for a normal company. How could Google's general counsel and its outside lawyers have permitted this to happen? The answer, analysts speculated, was that Larry and Sergey didn't really want their employees to know how much money the company was making, because the news would leak and competitors would find out. So despite the regulations, they had played it close to the vest.

Google's unusual auction method for selling stock in the IPO was not devoid of potential problems or risks either. "The auction process for our public offering may result in a phenomenon known as the 'winner's curse,' and, as a result, investors may experience significant losses," the company warned. "Successful bidders may conclude that they paid too much for our shares and could seek to immediately sell their shares to limit their losses should our stock price decline."

With the stock market entering the dog days of summer and the SEC still reviewing Google's unorthodox auction, speculation mounted that the search engine giant would do the logical thing and wait until after Labor Day to go public. It appeared the company was going to raise billions of dollars in the IPO, and nobody attempted this kind of unconventional deal in August, when Wall Street took a nap. It was a time when investment bankers fled to the Hamptons, Martha's Vineyard, and more exotic locales, since few decisions of consequence were made then. Major clients, including companies and investors, tended to vacation then too.

But Larry and Sergey were different. They found the entire public offering process to be enormously time-consuming and a wasteful distraction from the search engine. The sooner they could get the deal done, the sooner things would return to normal. So they pressed forward, even as issue after issue arose in the summer of 2004, giving Google the worst publicity it had experienced since its founding.

One could rationally argue that the combination of a weakly performing technology sector in the stock market, the summer hiatus, and the string of negative publicity that made Google appear poorly managed, made this just the kind of public offering that would do much better a few months later, in the fall, after the heat died down. Brin and Page, however, feared more damage to Google's reputation as Wall Street firms, competitors, and analysts found fault with the company's projections or its lofty price range of $110 to $135 per share, about 150 times its per-share earnings. The proposed price range drew criticism too, since it seemed bubble-like to those who remembered the technology meltdown several years earlier. And despite Google's efforts to reach out to individuals, its auction was complex. It required individuals to establish accounts at certain firms and follow a specific set of hard-to-decipher rules that were especially tricky for novices. The founders wanted to get the offering completed as soon as possible, and stop the bad-mouthing of the firm.

During those months in the summer of 2004, Google looked increasingly vulnerable. The IPO was being covered so closely that every error or potential risk was magnified. All of this undermined the original message from the founders that investors who wanted to own a stake in Google had to trust Larry and Sergey. Doing so seemed to be a riskier and riskier proposition. For a company that had managed its image to perfection, the warts began to make the Google Guys seem less like the Wizard of Oz and more like the man behind the curtain.

On Wall Street, some firms decided the deal was more trouble than it was worth. Merrill Lynch dropped out without public explanation, making its retail and institutional brokerage customers cautious about the stock. It was one factor, but not the only one, that eventually led some financial analysts and advisors to recommend to their customers that they watch from the sidelines whenever Google did go public, and then decide, after the firm's shares began trading, whether they did or did not want to invest. Fears also spread that in an overheated auction environment, those who purchased stock in the IPO could be paying top dollar, leaving the shares with nowhere to go but down.

But before that day came, fresh issues arose in the minds of investors about whether Google was ready for prime time as a public company.

With media reports pouring forth about SEC probes, other legal problems, and the risk of the auction leading to a sky-high price in the IPO, serious questions arose about whether Google could get the deal done at all. At a certain point, it didn't matter whether it was Google's rapid growth causing the problems, its failure to abide by every aspect of securities laws along the way, or a Wall Street campaign to sully the IPO. The mood had shifted since the April IPO filing. Then, the brand name had been beyond reproach. Now, questions swirled. Rumors abounded that shareholders, large and small, were not participating in the auction, and that demand for Google shares would be so low the company might have to pull the deal completely. Confidence in the company's management plunged. Some analysts blamed Morgan Stanley and Credit Suisse First Boston for urging Google to set the initial price range for the IPO too high. Whatever the cause, Google was on the retreat.

Just when it looked as though nothing else could go wrong with Google's IPO, *Playboy* magazine published a major interview with Larry and Sergey called "Google Guys." Under stock market rules, the interview was a potential violation of the quiet period. The *Playboy* fiasco raised questions anew about the competence and maturity of Google's leaders. After all, this wasn't exactly *The Wall Street Journal* or *BusinessWeek*, the traditional media venues for successful American businesses. Better known for its nude centerfolds and playmates than articles, *Playboy* had conducted the interview months earlier, in April, and released it amid the Google storm to garner maximum publicity.

While some people found it amusing, the company's major investors found it maddening. The SEC, already investigating Google for failing to register its shares, now had to decide whether to sanction the company for violating the quiet period. To answer the question, SEC lawyers would have to read the interview. Comedians and editorial cartoonists had a field day with the notion of government lawyers thumbing through the pages of *Playboy*, pretending not to look at the photo spreads, as they carefully read the Google interview. Could the Google Guys, who were stung by new revelations at every turn en route to going public, be trusted with billions of dollars of other people's money?

Playboy's contributing editor David Sheff, who conducted the interview, wrote that when he arrived at the Googleplex, "Brin was indeed having fun, playing a sweaty game of volleyball in an open-air plaza. Dragged in shoeless from the court, he contemplated questions with great seriousness while occasionally stabbing at a salad. Throughout our conversation he and Page, who wore shoes, rarely sat down. Instead they stood up, leaned on their chair backs, climbed on their chairs and wandered about the windowed conference room. It's apparently impossible to sit still when you're engaged in changing the world."

All in all, Wall Street was having a field day talking the deal down, and it no longer seemed likely, given weakness in technology stocks and tepid demand for Google shares, that the company could go public in the pricey $110 to $135 range announced by the investment firms. There were other issues that made potential investors uneasy too. The most serious roadblock was the ongoing legal dispute between Google and Yahoo over ad system patents.

In a bid to get the deal done, John Doerr and Michael Moritz, the firm's two key venture capitalists, asserted themselves and called for Google to settle the patent dispute with Yahoo. Google gave Yahoo 2.7 million shares to drop the litigation. The world would never know how much Google may have overreached in copying Overture's ad system. The settlement cost Google hundreds of millions of dollars in stock, but it lifted the uncertainty hanging over the IPO. "You can't look at this and decide it is a nuisance value settlement. It is, rather, a substantial acknowledgment by Google that it was violating the patent laws," said David Rammelt, an attorney opposing Google in a separate trademark case. In court filings, Google asserted that it had not violated the patents.

Given all the problems, the team considered putting off the IPO until sometime in the fall. "There was a brief discussion in August about the pros and cons of delaying it," Moritz said. "The question seriously reviewed was whether we would be better off to wait until October of 2004. The decision was to pursue it, to get it over with, and concentrate on the amazing and relentless demands of daily business activity. I thought that was the right thing to do. Otherwise, we would have had a three-month death march, which is a major diversion."

Since Sergey and Larry had no desire for the bruising public offering process to drag on a day longer than necessary, the company's fate was in the hands of the Securities and Exchange Commission in Washington. "The SEC is reading a lot of issues of *Playboy* right now," quipped Tom Taulli, author of a book about IPOs. Would the lawyers there conclude that Google's IPO had to be canceled or postponed?

Back in California, Google's lawyers at Wilson Sonsini endorsed a possible solution: file the *Playboy* interview as an appendix to the company's SEC registration statement, and make it part of the official material available to all investors prior to the public offering. The ploy adhered to the SEC's guiding principle that disclosure to investors solved many problems. And fortunately for Google, it worked. While reserving the right to investigate the *Playboy* matter later, SEC counsel determined that disclosing the story as part of the Google filing, and correcting inaccuracies in the article, would be sufficient to go forward.

Finally, Google could accept bids from investors, set a single auction price, and get the whole thing moving. But the $110 to $135 per share range proved too rich. All the damaging disclosures about Google, a 40 percent plunge in the price of other technology stocks, and the complex auction rules depressed the demand for shares. The battering Google had taken in the press and among the Street's chattering class hadn't helped either.

To shore up confidence in the IPO, Doerr and Moritz took additional steps: they cut the IPO price to the range of $85 to $95 per share—low enough to attract more demand, and a chance to restore confidence in Google, since the first day of trading would likely see an upward swing and make investors feel good about Google. And lest people view the August IPO as a desperation play, Doerr's firm, Kleiner Perkins, and Moritz's Sequoia Capital reversed course by keeping all of the Google stock they had planned to sell, a signal that the smart money anticipated Google's stock price would rise.

At long last, after more headlines than any public offering had had in years, the Google IPO, under the ticker symbol GOOG, went out at $85 a share on the NASDAQ exchange on August 19. The 19.6 million shares of stock initially could not begin trading at 9:30 A.M., when the market opened, because demand exceeded supply for the relatively small number of shares being offered. When trading did begin at 11:56 A.M., the stock jumped $15.01 to $100.01. The offering raised $1.67 billion and gave the company an initial market value of $23.1 billion. The Wall Street firms comanaging the deal, Credit Suisse First Boston and Morgan Stanley, received less than half of their usual fees.

Suddenly, Google had a stock market value higher than many older, well-established enterprises. Some investors who bought shares in the IPO quickly sold them for a tidy profit. The novel auction process, in the end, achieved one of its two goals: the company, not Wall Street, had remained in control by allocating shares equitably based on investor bids. This more egalitarian approach avoided the scandals Wall Street had been reeling from in recent years, with hot IPOs going to the favored few. What the auction failed to achieve was a dearer price for Google stock. By selling shares at $85, the company left plenty of money on the table. Had the offering gone more smoothly, the bidding process been simpler, or Google not been so determined to get the deal done in August, a higher initial public offering price would have been achieved. And the company would have been able to stash more money in its coffers.

On the day of the IPO, Sergey Brin showed up for work at the Googleplex, a sign of the company's ongoing focus on the day-to-day work at hand. His 31-year-old copresident and cofounder went to New York, where he joined CEO Eric Schmidt and venture capitalist John Doerr in ceremonies marking the opening of public trading. Page and Schmidt had breakfast with NASDAQ officials preceding the opening, where Page, despite becoming a billionaire on paper that very morning, appeared somewhat distant and removed from the action around him. "It will be interesting to see what happens," he said in response to overtures from the officials. Looking uncomfortable in a coat and tie, according to *GQ* magazine, Page managed to sit in a plateful of crème fraîche, temporarily soiling his posterior, which Googlers helped him wipe off. "These things happen," Schmidt said. "We've seen worse."

For Brin and Page, the march toward public ownership was finally over. They had done the IPO their way, effectively breaking a Wall Street cartel. They could restore their focus to the business of running their business. Still, with shareholders they had never met, the founders faced an entirely new level of public scrutiny and responsibility.

The Google IPO marked a watershed in the relationship between Silicon Valley and Wall Street. Larry and Sergey had pulled off one of the biggest initial public offerings ever, maintaining control over the process and earning the respect of corporate chieftains who had been through the Wall Street mill themselves. Speculation was rampant about whether the public offering heralded the dawn of a new era in deal-making for technology companies, or was merely the latest sign of Google's powerful and unique enterprise. Few other companies could have pulled off the August IPO under such duress. As for Larry and Sergey, they breathed a sigh of relief. They knew things would never be quite the same, given the legal and other issues hanging over the now public company. But at least life inside the Googleplex had a shot at returning to normal and being fun again. Days later, Larry and Sergey headed to Burning Man—a sign, friends said, that despite becoming billionaires, the Google Guys apparently hadn't changed.

CHAPTER 18
Charlie's Place

On the day that Google went public, Sergey and others at the Googleplex celebrated by devouring gallons of flavorful ice cream from Ben & Jerry's. Nobody had known what unusual treat awaited them at work to mark the occasion. Traditionally, companies would have had champagne, complete with corks popping and effusive speeches about the company's bright future. But Google did things differently. Sergey was as surprised as anyone to see the special all-day ice cream bar, since he had nothing to do with the decision. Instead, that call was made by Google's informal minister of cultural affairs. His name doesn't appear on the public filings listing the most senior executives, but his contributions to Google played an enormous role in shaping the company's enjoyable, nurturing, and productive environment. "The ice cream was fun, lighthearted in nature, not what anyone expected," said Google executive chef Charlie Ayers. It was, "'We gotta zig again because they think we are going to zag.' Everyone was very happy."

Back in 1998, when Google was less than a year old and had a dozen or so employees in a cramped downtown Palo Alto office, Sergey met with Charlie Ayers to talk about hiring him as the company's chef. Good, healthy, free meals for employees, Brin said, were going to set Google apart from other firms. Ayers had cooked for the Grateful Dead, a band famous for its cult following, and his association with the rockers would add panache to Google when it sought to hire engineers. But when Ayers met with Brin, the conversation seemed zany. There weren't enough employees at Google to justify hiring him. The company's future was unclear. And the cash-strapped technology start-up certainly didn't have money to burn. Their conversation was direct.

"I was like, 'Sergey, you need a *chef*?'" Ayers recalled.

"We are going to be tens of thousands of people," Brin said. "We need a chef."

"You don't even have a kitchen," said Ayers.

And with that, Ayers bid Brin farewell, strode out onto University Avenue, and returned unhappily to his job as a personal chef for a wealthy family that didn't appreciate him. He would have to find a different means of escape. But remarkably, eight months later, Ayers heard that Google was interviewing chefs and giving them tryouts. To attract candidates, Google had posted an ad on its Web site.

HEAD CHEF—THE GOOGLERS ARE HUNGRY!!

> One of Silicon Valley's hottest and fastest growing Internet companies is looking for an experienced and innovative gourmet Chef to manage all aspects of Google's on-site Cafe. In this position you will be responsible for managing the Cafe, from menu planning to final presentation. The experienced Chef of choice should be creative and healthy in planning menus for Googlers. Here's a group of people with well traveled refined palates with a craving for epicurial delights.
>
> *The only Chef job with stock options!*

By then, Google was some 45 employees strong and had moved its offices to Mountain View. Before Ayers went to check things out for himself, the company had given 25 chefs tryouts and rejected them all. Still miserable in his job, Ayers was open-minded and ready to make a move. At least Sergey and his employees seemed friendly and fun. And there was nowhere else for them to eat nearby except McDonald's and Krispy Kreme.

"I came in and looked at the place and there was this little antiquated electric kitchen," Ayers recalled. "I said, 'We can do this until we get a real kitchen.' They said, 'Don't worry. You will have the best kitchen that money can buy.' Sergey was adamant about the food service and keeping employees on campus, and keeping productivity up, and encouraging them to come in every day not knowing what they were going to be served. I wouldn't put out the menu until ten minutes before lunch. They wanted burgers, hot dogs, burritos. They were kids. I said, 'I was hired to do something else: really healthy, eclectic foods that were organic.' I sold them on the organic thing. The first time I met them, Sergey talked about saving the world and a lot of ideals. I said, 'This goes with what you want to do.' Everyone there only knew food from a customer standpoint. I knew what I was talking about. I was very ready to make a change and love a good challenge."

Hired on November 17, 1999, Charlie Ayers became Google employee number 56. Like many who joined the company early, he took a salary cut when he accepted the position. At 33, he was one of the company's oldest employees; most of the others were still in their twenties. Sergey and Larry wanted free, wholesome food for everyone. It was a perk with a purpose. It would keep people near one another and their desks; prevent them from developing poor eating habits that would diminish productivity; eliminate the time they would otherwise spend going out to lunch and worrying over plans; and create a sense of togetherness. Six months and thousands of meals later, Ayers was exhausted. He was feeding the Googlers and cleaning up by himself. "I said, 'You guys are killing me.' They said, 'You are only cooking for 50 people a day.' I said, 'Have *you* ever cooked and cleaned for 50 people a day?'"

Still, the smell of something special was in the air, and Chef

Charlie loved being a part of it. Employee motivation was sky-high, and everyone had a sense of purpose and teamwork. They were a hardworking family, and he was the one putting the food on the table. They appreciated him. Around the table, he played the part of the older uncle that Sergey, Larry, and others enjoyed talking to. He remembered their names. And as the company grew, Sergey kept his word, giving Charlie the freedom to hire additional cooks to assist in the kitchen. "I could feel the energy. They had it," Ayers recalled. "Everyone was so focused and into it, and they all had one goal: to make this company successful. It was 'Look at what we did,' not 'Look at me.' It was a total team effort. Once they said, 'We are going to have a wire party this weekend.' I didn't know what that was. I went to the data center in San Jose and found myself on the ground pulling cables. We all pitched in."

By the time Google moved into its permanent headquarters in Mountain View in January 2004, the company was known for providing fantastic, free meals for its employees, who were young, mostly single, and enjoying every bite. Breakfast was breakfast, but when it came to lunch, Charlie kept everyone guessing. The word around Silicon Valley was that the food at Google was better than the fare at area restaurants. It didn't matter if you were a vegetarian, or into Asian or Middle Eastern cuisine, or an engineer filling a utilitarian need to put something into your stomach so you could get back to your desk as fast as possible to keep writing code. Google had the best fresh food around, and you didn't have to think about where you were going to eat, who you were going to eat with, or whether you had any cash in your pocket to pay for it.

When Googlers returned to work on a Monday and told Charlie they had spent money eating out over the weekend and didn't enjoy the food as much as his, he told them why. "The difference," he said, "is because I care about you folks. I cook with love. Others cook because they want your money."

Before long, Charlie's Place, as the company's café was affectionately called, became an institution. And Charlie's influence on Google's reputation also grew, as word spread that a former cook for the Grateful Dead prepared gourmet meals for the hottest technology company around. With additional food, drinks, and snacks placed strategically around the campus, people felt cared for. In a Google employee survey asking what people liked about their jobs and coming to work, nine out of ten people cited the food.

After learning how much Charlie and his food meant to technology's finest, Google posted a sample menu on its Web site alongside the list of current job openings:

ONE OF OUR LUNCH MENUS . . . REALLY!

Soups

- Sweet Potato Jalapeno Bisque with corn
- Creamy Cauliflower Parmesan

Salads

- Warm Southern Chicken Salad tossed in a spicy buttermilk dressing with toasted pecans, corn, green onions and tomatoes
- Tortellini Primavera salad—organic tortellini mixed with organic zucchini, yellow squash, tomato, sweet peas, pesto vinaigrette
- Organic mixed greens

Entrees

- Grilled Petite New York Sirloins seasoned with Creole spices served with a Crescent City steak sauce and crispy organic onion rings
- Organic Tofu Mushroom Ragout—domestic and wild mushrooms, vegetable stock, leeks

The company also had a top-ten list of the best reasons to work at Google on its Web site that concluded with: "There is such a thing as a free lunch after all. In fact we have them every day: healthy, yummy, and made with love."

Charlie's influence at Google extended beyond mealtimes. On Friday afternoons at the Googleplex, everyone gathered for a "TGIF" beer, soda, and munchies bash, better than the ones they had had at Stanford. Once a month, TGIF was strictly social, and Charlie began hiring entertainment. On the other Fridays, Larry, Sergey, or Eric typically spoke about how things were going at Google and answered questions. They also used the occasion to introduce new employees, known at the company as "Nooglers." To warm things up for the Nooglers and the others too, Charlie took some surfboards, had them embossed with the Google logo, and loaded them with sushi, chocolate fondue, and other finger foods.

"My goal every day was to create the illusion you were not at work but on some type of cruise and resort, through the cuisine, décor, entertainment, and the extra things we did," Charlie said.

"There was electricity in the air. Everyone was on fire. As soon as you walked in, you were hit with this onslaught of color. Vibrant colors in the lobby, hues of primary colors, lava lamps, people riding around on scooters in the hallways, things you didn't see anywhere else. People had their dogs at work. You walked in and looked around and people wondered, 'What kind of place is this?' It was like an extension of Stanford in a lot of ways. Creativity was not squelched. We did a lot of creative things for food service. We had to erect a tent for two and a half years so we could have people dine under a tent, with a mobile catering truck in front. That was the kitchen extended.

"None of them," Charlie said of the young Googlers, "had ever experienced in the workplace or at all, the level of service and food I was giving them. I said, 'You are getting paid to try something new to eat.'"

Soon Google was posting fun facts on its Web site about the foods consumed at the Googleplex. Given the international mix of the company's employees, Ayers varied the menu, though always with high-energy foods that employees would metabolize well. Among other kinds, he said, "We did Southwestern, classic Italian, French, African, my version of Asian, and Indian." And he frequently used the Google search engine to research recipes.

"I helped them create a lot of the culture through the food we did there, the barbeques, and having live music. I hired all the live entertainment for them." On his four-year anniversary, they presented Charlie, who by then held the title of Executive Chef, with a T-shirt that read, "Serving Grateful Googlers since 1999."

Charlie's fried chicken was the best that Googlers had ever tasted. His extraordinary influence over Google culture and image grew mythic. He even posted the following on the company's Web site for millions around the world, and inside Google, to read in 2004.

> Long before cooking at Google, I worked in the kitchen of the Waldorf-Astoria Hotel with a Southern gent named Robert Brown. Story was that at one time in his life, Mr. Brown had cooked for Elvis Presley. Mr. Brown never gave us the details, but he did let us know that the King loved his fried chicken and biscuits.
>
> Mr. Brown had a fluffy white cloud of hair floating above a face the color of molasses, garnished with a big gold tooth protruding from his mouth and thick dark glasses he never took off, even in the kitchen. He was a primal cook—he couldn't tell you why he did what he did, but he knew when it was "GOOOOOD." It was better than good. It was the best southern fried chicken I had ever tasted, and still is.
>
> One day I got up the courage to ask him for the recipe, and he told me, "Charlie, I normally don't give out my prize recipes, but you, boy, have got the touch. And none of my boys are in the business, so I will give it to you." The secret to his fried chicken was marinating it in buttermilk for a long time, and adding just about every damn spice he had on hand.

Whenever I serve it today, I can hear Robert Brown saying, "Charlie, you make this chicken for people, you'll be making friends for life." And I hope I have. Here's the Google-sized recipe.

BUTTERMILK FRIED CHICKEN ELVIS LOVED

½ c thyme
¼ c oregano
¼ c basil
½ c onion powder
½ c garlic powder
½ c dry mustard
½ c paprika
¼ c chili powder
½ c celery seed
2 Tbsp salt
½ c coriander
½ c cumin
⅓ c kosher salt
¼ c cayenne pepper
½ c ground black pepper
¼ c ground white pepper

3 gals. buttermilk
3 cases organic free-range chicken (roughly 30 chickens, divided into 1.5- to 2-lb. sections)

Mix these amounts of the dry ingredients together in a large bowl, then whisk in the buttermilk until it's thoroughly mixed. Pour the batter over the chickens and marinate for up to five days—keep refrigerated, of course.

For frying
Now mix another 4x the above dry ingredients, and add:
2 lbs. cornstarch
8 qts. all-purpose organic whole wheat flour

Dredge the marinated chicken pieces in the dry herbs/flour/cornstarch mixture. Fry the dredged chicken in a large skillet with hot peanut oil @ 375 degrees. Once chicken has reached a golden brown color, finish cooking it in the oven.

When Larry and Sergey had their thirtieth birthday parties, they turned to Charlie to plan the menus and prepare the meals. He was honored to have been asked. And he knew their favorite foods. For Larry's birthday bash—a giant gathering, since invitations went to every employee of Google—he stuck with simple foods. He made fresh hoagie sandwiches and pizza, and for dessert, a butter-cream sheet cake decorated in the Google logo's primary colors. "People were saying this is nothing special," Charlie recalled. "I just made foods I knew he liked. This was not Quiznos or Dominos. It was special."

Sergey's thirtieth birthday party was smaller and more exclusive, and the food reflected his eclectic tastes. Charlie prepared sushi, different types of Indian appetizers, Mediterranean appetizers, and other round-the-world specialties. Instead of heavy meals with big courses, Sergey preferred finger foods, in part because they made it easier to carry on conversations. "One or two bites—nothing that was going to distract," Charlie said. "They are very intent on conversation and networking." Given the mix of foods, Sergey's party was "a little flashier," enabling the executive chef to have more fun. Dessert included chocolate-covered strawberries injected with Grand Marnier; truffles; mini-apple croustades, and baklava. "They had really cool entertainment. They had a magician and a contortionist. He had fun."

Larry and Sergey recognized the important contribution Charlie was making to Google by increasing his salary and giving him the chance to buy cheap stock options when the company was still private. Meanwhile, Charlie was starting to get offers from outside the company to open his own restaurants. Investors were offering to back him. But he didn't want to leave Google; his work wasn't complete. Charlie didn't have the cash to buy the stock options, so he turned to his father for a loan. Charlie knew a lot about food and had enjoyed cooking ever since he was a kid, but he didn't know much about stocks and bonds. His father knew even less. "I shared a desk with an engineer and a guy in finance and marketing. It was all new to me. I asked all kinds of questions. They said, 'Be a smart guy and buy your stock options.' My father said, 'It is a scam. Don't give them your money.' He lent me a small amount of cash to buy the options I was offered."

Charlie said the notion that he had been the longtime chef for the Grateful Dead was an exaggeration by the media that Google did not correct. Nevertheless, it took on a special significance inside the company that bonded him with Larry Page. Charlie was a friend of the chef for the Dead and helped out one night when someone else couldn't make it. "He liked my work and continued to give me calls. You didn't get paid but got backstage passes." Charlie said he pitched in only occasionally when the group was performing in the Bay Area. It was exciting, reminiscent of the feeling that Charlie, who grew up in Brooklyn and New Jersey, had with his first restaurant job in ninth grade. He fell in love with the profession and concluded he would always have a job, since people need to eat.

One evening at Google, Charlie was making dinner for Larry, his brother Carl Jr., and their uncle when he noticed the uncle bopping when he heard Grateful Dead tunes playing in the kitchen. Charlie was curious, but didn't want to pry. The uncle smiled at Charlie and then told him about the emotions and memories the music conjured up.

"I wanted to let you know that when Larry hears this music, he thinks of his father," the uncle said.

"Why?" Charlie asked.

"Larry never told you? We used to take him to Grateful Dead concerts when he was a kid."

"When he talks to you," Carl Jr. told Charlie, "you remind him of Dad."

Charlie said that hearing about these feelings, and learning that Larry had joined his brother and mother in a peace march in Oregon opposing the war in Iraq, changed the way he perceived Google's copresident. "It helped me understand him. After that I—I did not think of him as just a scientist."

Several months after the IPO, Charlie Ayers made the difficult decision to leave Google to open some restaurants of his own in northern California. As Google stock shot up in price, his stock options had become valuable, providing the money that he and his wife needed to buy a home. Other Googlers used their stock proceeds to buy new BMWs, Mercedeses, and Porsches, which started showing up in the company's parking lot. Things had started to change at Google after it went public and continued to grow at an extraordinary rate. For example, there were an increasing number of employees who had never met or talked one-on-one with Larry and Sergey. And Ayers had also observed some new internal dynamics.

"Some department heads became more proactive to meet their numbers and in insisting that their lead managers and directors were bringing numbers in and staying on course, and not playing around like it was pre-IPO," he said. "It is really important, even if the company is making money hand-over-fist and reporting numbers beyond Wall Street projections, to run the business like every dime counts."

In food and certain other areas, Ayers said, Google, under the tutelage of Larry Page, typically insisted on a 35 percent discount to enter into long-term agreements with suppliers. It was a very effective way to hold down spending. While some vendors initially balked, they often came around since Google was spending so much money. It was a reliable way for them to cover their overhead, ensure they had plenty of demand to keep operating, and position themselves to profit from sales to other companies.

Ayers said another reason he decided to leave Google was to get better control of his life. The intensity of the job had left him out of shape, graying, and longing for the healthier lifestyle he could now afford. He was not alone. "Some who left gave up X amount of years or have families. A lot of female engineers who helped start the company wanted to become moms. People left for health and sanity issues. I have put myself on a diet and exercise program since I left Google. I was starting to look like chefs I used to work for. I'm not an old guy, so I want to preserve myself."

In 2005, after five and a half years of being an integral part of the fabric of Google, Ayers notified Larry and Sergey of his plans to leave the company by writing a formal letter of resignation and putting it on their computer monitors. "They took it really hard," he said. Page regarded Ayers's accomplishments as significant. "Charlie has single-handedly turned around the food service culture in Silicon Valley by doing what he knew was best—making people comfortable and happy," Page said.

Ayers considered himself to be extraordinarily lucky to have landed at the Googleplex at just the right moment to be part of something that mattered, and to benefit immensely from stock options. "It is very rare that people in my business get this opportunity," he said. "I got very lucky, cherish what I have, and am going to hold on to it."

At one of the last Friday afternoon TGIFs that Ayers attended, Larry was away in Africa but Eric Schmidt asked Charlie to come up onstage. There were 5,000 T-shirts given out to Googlers that had Charlie's face atop the body of Elvis Presley. As he had been advised, the fried chicken that Elvis loved had brought Charlie many friends. He autographed T-shirts for hours. "Sergey and Eric gave me a big hug and people started rushing the stage and hugging me. There was a long standing ovation and people were crying and taking pictures. Thomas Friedman and Robin Williams happened to be there that day. I got off the stage and they were asking me for my autograph. Eric said to Robin Williams, 'Charlie created the culture here for us.'"

CHAPTER 19

Space Race

Flying to Spain in the fall of 2004, Sergey Brin and Larry Page received some bad news. Omid Kordestani, head of worldwide sales, notified his bosses that Yahoo had beaten Google in a competition to be the exclusive provider of ads for AOL's European Internet service. Brin, who badly wanted the AOL business in Europe, was determined to see what he could do to reverse the situation. An earlier deal between Google and AOL had sent the search engine's presence in the U.S. soaring, and it needed that sort of momentum in Europe too. For several years, Yahoo had been providing ads to AOL's 6.3 million computer users in England, France, and Germany, as well as to various AOL-affiliated Web sites. And in this latest competition Yahoo had bested Google very simply: it had offered more money and better terms. Without asking too many questions, Brin sprang into action, giving his man on the ground, Kordestani, a forceful directive: notify Philip Rowley, head of AOL Europe, that Google's founders were diverting their flight and would land in London instead. They wanted to meet with him personally, and they wanted to do it that day. He also instructed Kordestani to raise Google's offer for the AOL Europe business.

Rowley flatly turned Kordestani and Google down, telling him that AOL had made a deal with Yahoo. "That is the way it is going to go," Rowley said. "The process is done."

Sergey Brin refused to take no for an answer. From 30,000 feet, he ordered Kordestani to guarantee Rowley that he and AOL would benefit enormously from meeting with him and Page, and to wait before signing any deal with Yahoo. He added that it meant an enormous amount to him personally to be given the chance to talk with Rowley himself. Their flight was en route to London, he added, and he wanted to know where and when to meet Rowley later that day.

Rowley, unsure how to proceed, contacted America Online headquarters in Virginia, next to Dulles International Airport. There, Chief Executive Officer Jon Miller, who had signed off on the Yahoo deal, got word from Rowley that Google's top brass were on their way to London to meet with him. Rowley told Miller that Brin and Page had sent a message saying that they placed a high personal priority on winning the AOL Europe business. Miller, a bright, highly analytical executive, knew the guys and admired Google greatly. But he and Rowley had a problem. They already had notified Yahoo that it had won the competition for AOL Europe. Miller decided that if the Google Guys came to London and merely topped Yahoo's latest offer slightly, AOL would stick by the Yahoo deal. If, on the other hand, Google took things to totally new heights, then the fair thing to do, since no documents had been signed, would be to reopen the process and let Yahoo know about Google's surprise bid. Rowley decided to arrange a confidential, off-site meeting with Brin and Page, and keep Miller in the loop.

Competitive and driven, Brin enjoyed diverting the private jet and going after the deal. He had the charisma, persuasive personality, and innate skill to make business deals happen. And he didn't want to miss out on this opportunity to power Google's global growth without giving it his best shot. As senior executives and Google's biggest shareholders, Brin and Page had the authority to make enormous, long-term financial commitments that Kordestani could not. At that moment, nothing mattered more to Brin than getting the Google brand name in front of as many Europeans as possible. As they flew toward London and reviewed the situation, Brin and Page decided that they would listen carefully, try to make Rowley an offer he couldn't refuse, and hope that they weren't showing up too late to make a difference.

Rowley sent word through Kordestani to meet him in London at the Milestone Hotel, near AOL's office. He did not want to arouse suspicion by having Brin and Page come to his London office suite. After the high-profile Google IPO, too many people could recognize them and start asking questions. From Brin's vantage point, at least Rowley was giving them the opportunity to be heard.

In a room at the hotel, Brin and Page met with Rowley and quickly put an irresistible, lucrative, risk-free offer on the table. It included a financial guarantee of tens of millions of dollars, higher than anything Google or Yahoo had bid previously. There was no need to haggle. Rowley summoned a small AOL deal team to the hotel, and Brin and Page called in their own deal guys. As the leaders of Google and AOL Europe met in one room, their teams pounded out the details of a proposed multiyear contract in another room. Whenever a question or issue arose, a member of one or both teams would check in with the bosses, and it would be resolved on the spot. With things moving fluidly, Rowley could see that Brin and Page were personally committed to winning AOL's business, so he excused himself and called Miller at the company's headquarters in Virginia.

Miller, who had been looking for ways to boost AOL's business prospects, was stunned by the surprisingly good news from London. After hanging up with Rowley, he walked down the hallway to find his senior aide, John Buckley.

"Something interesting is going on in Europe," Miller told Buckley. "Google has come in with an offer that is significantly better than Yahoo's. The right thing is to let Yahoo get in the game."

Miller, who was in the midst of discussions with Yahoo about various matters with the implicit understanding that Yahoo had won the AOL business, now wanted AOL to inform Yahoo executives about the potential changes in the deal. Rowley contacted Yahoo immediately and told executives there that Google had substantially topped its bid with a financial guarantee, so AOL had reopened the process and would be willing to entertain a revised offer from Yahoo. There was a very big gap between the two offers, Rowley told Yahoo. A short time later, Yahoo officials, miffed over the way things had been handled and upset about the direction they were heading, informed AOL that they were not going to get into a bidding war for the deal.

Brin and Page were confident that the high price they were paying would prove beneficial in the long run as they won the loyalty of computer users across Europe during a crucial period of Internet expansion. They signed a binding agreement that evening. To make it official, someone from AOL had gone to the bank earlier to get 120 British pounds in cash. Under British law, some consideration, or cash, had to change hands to make an agreement binding, and it did.

"Their physical presence made a big difference," Rowley said of Brin and Page. "You want to be able to trust the people you are going to work with. We were just getting to know Google from the European side of AOL, and for them to come over was really, really helpful. What we couldn't afford to do was let Google come in and not get a lockdown agreement. It had to be done fast if they were serious."

Rowley said Brin and Page, who had flown all night and then negotiated a deal, were wiped out. Still, he said, "they were pretty laid-back and cool. I wanted to do business with these guys. If they hadn't been there, this would never have happened."

AOL's Miller said that Brin and Page demonstrated that they not only were visionary founders of a business and technologists but also hands-on managers and aggressive businessmen. It was unusual to find engineers with their persistence and deal-making skills. This, he said, is a distinguishing feature that sets Google apart from many other companies. With Brin, Page, and Eric Schmidt at the helm providing extremely strong leadership,

Google's prospects look promising and they are a great partner for AOL, Miller said, adding that Brin in particular forced the AOL Europe deal to change course.

"The founders and Eric are very much operating stewards of the business," he went on. "They take the business seriously and engage in it directly. They are good guys. The three of them seem to find a way to work together well. AOL and I have tremendous respect for what they have achieved and how they have done it. Not everyone has done so well. No one else has done it so fast. And few stay as true to the founding principles as they have."

On October 19, exactly two months after Google's IPO, the search engine and AOL issued press releases announcing their expanded partnership in Europe. It was a bitter defeat for Yahoo's Overture division, the second major loss to Google for AOL's business alone. "It's a competitive market. Our partnership with Overture has been successful, but now we have chosen to go with Google," said a spokesman for AOL Europe.

Yahoo's Overture division said it chose to walk away rather than make a deal with AOL Europe that did not make financial sense. "Overture engages in distribution partnerships that make strategic and financial sense for our business," the company said. "In the case of AOL Europe, these criteria could not be met."

There was no mention in the Google press statement of the role played by Brin and Page. Kordestani, the sales executive, said the deal was a milestone for both companies. "AOL's European operations will benefit from the revenue opportunity, while its users will enjoy an enhanced online experience through the addition of relevant commercial information."

The pace of innovation at Google in the aftermath of its IPO and its victory over Yahoo in Europe proved unnerving for competitors. With new features and products constantly flying out of the Googleplex to millions of users around the world, it was as if Google had engaged competitors in a space race to see who could amass the biggest and most buzz-worthy arsenal of features. Nobody else got Google's press coverage, so every move it made was flagged for the world to see. And now that it was a public company, it attracted even greater attention from the business press than it had when private, since many financial media outlets and analysts covered only publicly traded firms.

With the hassles of the stock offering behind it, Google's new public platform freed the company to do what it did best: focus on serving users through constant innovation in search and related offerings. Google not only triumphed in the space race during this period, leaving Microsoft and others to play catch-up on several fronts, but it also expanded its lead in its core search and advertising businesses in the U.S., Europe, and Asia.

Industry veterans noted qualities about Google's approach that distinguished it in major ways. For example, its work was not motivated by fighting with a business adversary and it avoided the temptation to create an archrival. "Google is not anti-anybody," said Jon Miller of AOL. "Most companies need a business enemy, and that is how they motivate themselves." Brin and Page, on the other hand, "are motivated by their mission. Clearly, they think very differently and are driven by their vision and business goals."

To keep Google moving, avoid bottlenecks, and hit targets of opportunity, Larry, Sergey, and Eric divided their work as leaders in ways that their job titles, or the unusual triumvirate structure at the top, didn't convey. They communicated with one another constantly, but they also established processes and routines, as well as clearly delineated responsibilities, that each maintained himself. Both Larry and Sergey carried the title of copresident and Eric wore the badge of CEO, but in reality, each had dominion over different functional areas, and shared responsibility over others.

That was not all they shared. Larry and Sergey inhabited a single, rectangular work space in the corner of the newly renovated Building 43 on the Google campus. There they worked side by side under the same roof as an army of computer engineers and technologists who also shared offices. The nondescript exterior of Office 211 didn't call attention either to itself or to the company's founders, and finding it required navigating a bit of a maze. In a version of upstairs/downstairs, Larry and Sergey sometimes used the main office space jointly, and at other times gravitated to an open-air office atop 211 that provided a view of the comings and goings of Googlers and their visitors. Though Eric had had an office right beside the guys when he first arrived, that was no longer the case. When the Google Guys moved into the dynamic new building—which Larry had helped design—Eric's office was now located across the way, in a building connected by a walking bridge.

Initially, when Larry and Sergey founded the company in 1998, Larry held the CEO title and Sergey was president and chairman. That arrangement had been determined by the flip of a coin. In connection with going public, Larry took on the title of President-Products and Sergey became President-Technology. On a day-to-day basis, Larry's work was more detail-oriented and hands-on, while Sergey's emphasis was on corporate culture, motivation, cutting deals, and monitoring projects with significant long-term potential.

As CEO, Eric oversaw operations, a mammoth management task in a company growing so rapidly. He focused on building internal accounting, financial, and other systems that scaled well as the company grew, and on establishing the special infrastructure necessary for global expansion. He also kept close tabs on Google's financial performance. It fell to Eric to make certain that various operating schedules were established, that deadlines were met, and that directors on the company's board of directors were consulted when necessary.

Each week, the executives held a "GPS" meeting—the abbreviation stood for Google Product Strategy. During these sessions, they heard proposals for projects and requests for resources. Often decisions were made on the spot, provided two of the three in the triumvirate agreed. Larry and Sergey sat side by side in these tightly run sessions. After years of working together, the two could quickly communicate on multiple levels through gestures and expressions.

As the AOL deal had exemplified, Sergey was skilled at establishing relationships and handling transactions of all kinds between Google and other entities. He had a knack for dealmaking, identifying what mattered most and then making things happen. His personality, keen intellect, and sense of humor facilitated those deals, as he knew how to smooth the inevitable bumps and lead the parties to solutions. He was also an arbiter and keeper of Google's culture, ensuring that the company had designated private spaces for mothers to nurse, free home food delivery for new parents, and an appealing, comfortable, healthy working environment. He constantly sought ways to embrace and motivate employees and maintain Google's campus-like image and its status as an extremely desirable place to work. Sergey also devoted energy to computer science issues and has been the catalyst for long-term technical research that could lead to scientific breakthroughs.

Stanford Professor Terry Winograd says Sergey has led the way on three Ps: policy, politics, and people. (When asked once what the motto Don't Be Evil meant, CEO Eric Schmidt famously replied that evil is whatever Sergey says is evil.) Larry, in contrast, was by nature drawn into the details of products and data center technology, including engineering issues of cost and performance. The two frequently had overlapping interests, however, and constantly debated and exchanged ideas.

Larry's primary focus has been on Google searchers, and users of its various other offerings. The product portfolio that attracted users to Google is something he worked on actively and constantly. He also played a gigantic role in hiring and recruiting employees, from executive talent to technical wizards. That didn't mean he met most new hires, but it did mean that he looked at résumés and reviewed selections made by others. Every Tuesday morning Larry studied, and then sometimes questioned, the hiring decisions that had been made at multihour meetings the previous day, which he had not attended. The hiring files he received were extremely detailed. For example, by the time an engineering offer made it to Larry's inbox, a job applicant typically had undergone a dozen individual interviews, taken tests to measure technical proficiency, submitted standardized test scores and transcripts, written software code on demand, and had his or her personality characteristics assessed by interviewers from across the company.

Larry also played a leadership role in prioritizing near-term technical research. Using a system known as the Top 100, he stayed on top of the 100 most compelling new and unfolding projects that might need staffing and other resources. Through this process, he helped to identify the most promising 20 percent projects.

When it came to Google's physical facilities, Larry played the dominant role in the company and got passionately involved in design and detail. Walking through offices of other companies, he would sometimes pull out a camera, as he did at BBC headquarters outside London, when he saw intriguing physical spaces that he wanted to remember. He has been the leading advocate for multiperson offices that encourage interaction, rather than cubicles that isolate people; and he also paid attention to special-purpose conference rooms and other spaces for teams to gather. He was uncompromising on the need for environmentally friendly practices and materials, including the lighting fixtures in the Googleplex.

When Building 43 was refurbished in 2004 to make room for Google's expanding workforce, Larry was in the trenches of the design and execution. The new building at the heart of the Googleplex is a playground for engineers that more nearly resembles a movie set than real life. "We wanted to be an engineering company, not a business or sales firm," he said. In redoing the facility, Google gutted the interior and created airy, loft-like spaces with mezzanines and alcoves, exposed ductwork, and a range of textures and metals that give it a striking industrial aesthetic. Bathrooms have extravagant, touchpad-controlled toilets with six levels of heat for the seat and automated washing, drying, and flushing without the need for toilet paper. The meeting spaces, outfitted in Googley color schemes and funky furniture, are part boardroom, part *Romper Room.* One conference space, perched by itself at the top of the building's grand central staircase, has a trompe l'oeil, or visual pun: giant floor-to-ceiling doors that seemed to open into empty air. (Panes of glass prevent anyone from mistakenly walking out.) Room names also run to the exotic: on the second floor, for example, meeting spaces are named after far-flung African cities like Rabat, Timbuktu, and Mogadishu.

Larry was determined to avoid creating a cubicle farm for the main workspaces in Building 43. Though some sections of each floor do have cubicles, their most noticeable features are huge flat-screen monitors, often two, even three abreast, that create the effect of the occupants having personal IMAX screens. Building 43 also has many tent-like offices, partitioned with a white plastic fabric and individually climate-controlled by ducts that feed directly into them from above.

In contrast, Office 211, where the founders reside, has real walls and two sets of sliding glass doors. Inside, plants, an air purifier, and monitors galore are near a pair of desks and a couch. Both Larry and Sergey work on multiple monitors displaying different information simultaneously. The open-air meeting area above 211, carpeted in Astroturf and outfitted with an electric massage chair imported from Japan, is strictly reserved for Larry, Sergey, and invited guests. A long foyer at the foot of the building's central staircase boasts a gargantuan whiteboard, covered with colorful scribbles about projects and technologies. It carries the label "Google's Master Plan."

While Google was churning out new features and products from the Googleplex and winning new users through free, word-of-mouth marketing, Microsoft—which had tens of billions of dollars in the bank—churned out new multimillion-dollar antitrust settlements with company after company. It was part of the continuing fallout from the software giant's tangle with the antitrust division of the U.S. Justice Department a few years before.

Microsoft Chairman Bill Gates talked often about what his company would deliver in future versions of Windows, and how it would eventually crush and trump Google as it had others before. But the Justice Department's antitrust lawsuit, filed the same year that Google was founded in 1998, had hung over the company and distracted senior Microsoft management as Google seized the Internet. "Over the last few years, we have been focused on resolving our disputes with other companies," said Microsoft general counsel Brad Smith. (The company's major settlements included a $1.6 billion payment to Sun Microsystems; a $775 million payment to IBM; a $700 million payment to Time Warner; and a $536 million payment to Novell Inc.) As Google's stature grew globally, Gates obsessed over how Microsoft was going to catch up and surpass the search engine, making it look passé, just as it had done years earlier with Netscape. The problem he faced was that with each passing day, Google's lead over Microsoft in search kept expanding. Google had global brand awareness, sound corporate partnerships, a worldwide sales presence, and other advantages—including unparalleled computer infrastructure—that would make it impossible for any overnight sneak attacks. And because Google was free to computer users, there was no possibility of a Microsoft assault based on slashing prices.

The more Gates talked, the less realistic he sounded. In the fall of 2004, Microsoft secretly planned a major announcement concerning a search engine of its own in test form. The headline-grabbing news it wanted to unveil was that its search engine had crawled five billion documents on the Internet, besting the comprehensiveness of Google's index, which had four billion. But a few hours before Microsoft released news about its purported triumph, Google announced that it had doubled its own index to eight billion Web pages, encompassing virtually the entire known Web and surpassing anything anyone else had done. Google posted the new page count on its homepage, in the spirit of the McDonald's signs showing the number of "Billions Served." Microsoft, caught flat-footed and red-faced, was once again seen as a laggard in a race with Google.

Still, Google's need for innovation remained immense. For one, it didn't have the so called "lock-in" that its competitors did. It did not ask users of its search engine to register, nor did it have services that caused people to spend long periods of time on its Web site. While AOL, Yahoo, and other Web leaders touted the growing amount of time users stayed on their sites, Google talked about how quickly it delivered search results, met users' needs, and moved them from its site to whatever they were seeking online. Some analysts saw this phenomenon as Google's great weakness. In contrast to AOL or Yahoo, which had plenty of chances to deliver ads to users hooked in through email accounts, and entertainment or business features, Google's quick turnaround time limited its ability to display ads.

What people didn't understand about Google, however, was that the potent Googleware blend of software and hardware gave the company more computing power than anyone else. It would not be easy for a newcomer to match its raw scale, said Peter Norvig, director of search quality. "We're like Dell," he said, referring to the way Google assembled and customized each of the personal computers in its vast network. Moreover, prior to each new product rollout, Brin and Page scrutinized the product's potential for rapid scaling up so that they could maintain a competitive advantage through size and distribution even as others mimicked and matched the ideas.

A fusillade of new releases kicked off in October 2004 with something close to Google's core search mission: desktop search. It was a fast, free, and easy way for people to find information of all kinds stored on their own computers as quickly as they could search the Internet. Microsoft had been promising something like it for years, but had yet to deliver. The Google innovation closed the gap between the accuracy and speed of searching the Internet and the maddening hunt for material saved on personal computers, from text documents to spreadsheets to long-lost emails. The downloadable program was made available in English, Chinese, French, and a half-dozen other languages.

"Considering how important the information on your computer is, it's always been a bit strange that you could find what you were looking for more easily if it were hidden on a Web site than in a corner of the hard drive sitting right in front of you," said Marissa Mayer, director of consumer Web products. "Google Desktop Search is like a photographic memory for your computer," searching and finding files in an instant from Microsoft Word, Outlook, PowerPoint, Excel, and AOL Instant Messenger.

The major potential negative of Google Desktop Search was that its use at work could enable nosy colleagues to swiftly find anything saved on someone else's computer. This made some people uneasy, but it was also a testament to how well the product worked. Google responded to this concern by building security features into the program, enabling users to turn it on and off, and providing them with a choice about whether to make all, or only some, of a computer's stored data available for search.

In addition to being useful for finding old items and those to come, Google Desktop Search embarrassed Microsoft, because it was giving millions of computer users the best way to find lost or misplaced files stored using Microsoft's programs. The feature also removed the burden of organizing material into folders and directories, since Google would retrieve it no matter how or where users stored it. Next, Google extended its reach into Microsoft's domain of the workplace, with a moderately priced way for businesses to index and search up to 100,000 internal documents and records. Packaged in a striking blue box emblazoned with the words "Google Mini," the appliance initially sold for $2,995 as Google sought to build off its consumer success by tapping the lucrative business and government markets.

Google remained synonymous with search while Yahoo offered a broader menu of choices for users. Both made announcement after announcement about innovations and products, each constantly seeking to upstage the other. After Yahoo claimed its search engine canvassed billions of pages more than Google's, they had a public spat about whether Yahoo was exaggerating. Some of the new features had obvious appeal; others seemed more cool than utilitarian, but served to keep both companies' names out in front of a broad range of potential employees and users.

Google pressed hard in a bid to stay out in front. It released satellite mapping and navigation services; fresh local search capabilities; ways for users to save personal search histories and build on them; video search based on the closed-captioning of television programs; mobile search by cell phone, BlackBerry, and other devices; the Google Labs Aptitude Test (GLAT), a mock standardized test for geeks and other potential Google recruits, containing a blend of complex and comical questions, including a small empty box with an invitation to "improve upon this space"; Google Suggest, a way for the search engine to propose search topics as you typed; and Google Scholar, a means for searching scientific journal articles, abstracts, technical reports, and Ph.D. theses. The company was smartly cultivating users from all walks of life, from ordinary users to business managers to university scientists.

When Google needed to convey an important message to its millions of users, its spartan white homepage made a handy billboard, since any bit of extra text instantly grabbed attention. Occasionally, the company posted new product announcements for a day or two, but in late December 2004, after a devastating tsunami wreaked havoc in Southeast Asia, it used the space to spread word of international relief efforts. This unique touch of humanity breathed life into Google, gave users an emotional connection, and enabled charitable organizations and aid programs to jumpstart worldwide campaigns to raise money and assistance for the victims.

The Google buzz attracted political luminaries too. From former presidents Jimmy Carter and Bill Clinton to former secretaries of state Madeleine Albright and Colin Powell, the search engine juggernaut had avid, high-profile users who visited the Googleplex to see firsthand what it was all about. Former vice president Albert Gore Jr.—who advised the company on international issues and launched a cable TV channel with their help—listed on his résumé that he was an unpaid consultant to Google.

With so many images and moments to capture, and digital photo use exploding worldwide, Google rolled out a new photo-search and storage system, with sophisticated, easy-to-use editing, captioning, and other features. Back in the realm of its core business, Google's popular Image Search grew to more than 1.2 billion images, the most comprehensive visual offering of its kind. Google also released quick new ways for computer users to search for stock quotes, taxis, and weather conditions. And in a stunning display of technology, the company released Google Earth, enabling computer users, from the comfort of their desktops, to visually fly to any place on the planet that they designated, with 3-D views along the way. Even in a culture of dramatic special effects in motion pictures and on television, this tool redefined the boundaries of search, converting individual computer users into explorers.

"There are times when even a hardened skeptic has to admit to amazement and delight at the sheer coolness of some of the things you can do on a personal computer today," wrote *The Wall Street Journal*'s Walt Mossberg. "And one of those 'wow' moments happens the first time you run a new program called Google Earth. It's an amazing demo and a great example of how much power computers and the Internet have put into the hands of average people. You may not use Google Earth every day, but it's worth fooling around with just because it's cool." Shortly thereafter, Google added a way to explore the surface of the moon at moon.google.com.

Sergey Brin and Larry Page were pleased. Googlers and their

families were, too, especially when Brin and Page rented out a nearby movie theatre for 24 hours and gave them all free tickets to see the latest *Star Wars* movie on the day it was released. With products and perks like these, Google was winning this round of the space race hands down.

CHAPTER 20

A Legal Showdown

Four months after snubbing Wall Street with its unorthodox initial public offering, Google posed another threat to the nation's business and legal establishment. The venue was the federal courthouse in Alexandria, Virginia, just across the Potomac River from Washington. Inside the stately building known in legal circles for its "rocket docket" and its handling of the most sensitive CIA cases, U.S. District Court Judge Leonie M. Brinkema presided with a sharp wit over her fifth-floor courtroom. It was the place where she had heard the celebrated terrorism case of Zacarias Moussaoui, the only person charged in the U.S. over the 9/11 attacks. This time around, at precisely 10 A.M. on December 13, 2004, the bailiff intoned "All rise," and the judge entered and rapped her gavel, calling to order all parties in *Government Employees Insurance Co. (Geico)* v. *Google Inc.*

The case lacked the global import and theatrics of the Moussaoui matter, where the defendant had jousted with Judge Brinkema by representing himself and trying her patience for days. But for the denizens of the growing Google Economy, the thousands of people around the world whose livelihood was tied inextricably to Google's continuing growth, the case was of paramount significance.

Geico and Google had little in common. Headquartered on the

border between Washington DC and Chevy Chase, Maryland, Geico was one of the nation's richest, most conservative, and well-established automobile insurance companies. Founded in 1936 by Leo and Lillian Goodwin, its niche was direct marketing to government employees, military officials, and affluent drivers, all of whom tended to have fewer accidents than the population at large, enabling the firm to offer lower auto insurance rates than competitors. The giant insurer had long been one of billionaire Warren Buffett's favorite investments, and in the mid-1990s he bought the entire company. Not surprisingly, Geico was represented in court that day by the old-line Washington law firm of Arnold & Porter.

Google, in contrast, was the upstart, the more aggressive and creative six-year-old enterprise based in California that took risks Geico would never dream of taking, or for that matter, even insuring. Rather than a staid Washington law firm, Google's lead counsel was Michael H. Page (no relation to Larry), who hailed from San Francisco, where he had earned a reputation for representing the underdog and won accolades for successfully litigating against movie studios and record labels over the issue of person-to-person file-sharing.

Geico had taken Google to court over an issue that reflected a subtle but important distinction in the way that Google and its chief rival, Yahoo Inc., operated when it came to advertisers. Geico objected to the way Google profited by selling ads to the insurance company's competitors that were linked to trademarked names that Geico owned. For example, another auto insurer could run an ad on Google by bidding on the word "Geico" or the trademarked product name "Geico Direct."

From Geico's vantage point, Google's approach was causing it to lose business unfairly. Geico alleged that the approach created confusion in the minds of consumers, who would type "Geico" into the search engine and then see ads for one or more of its competitors. To Geico, this was both blatantly misleading and wrong, trampling on the legal rights it had to protect its brand names, which were registered with the U.S. Patent and Trademark Office.

And it also meant that other insurance companies piggybacked on Geico's hefty radio and television advertising budget, since the company spent millions promoting the names "Geico" and "Geico Direct," only to have competitors allegedly steal business away by playing off the well-known name. Yahoo was less aggressive than Google, and Geico dropped its preliminary legal action against Yahoo after learning that the search engine refused to allow companies to post ads based on trademark-protected names of others. If any ads did slip through, Yahoo took them down as soon as someone complained.

Google's policy on trademarked names, once fairly strict, had become more permissive in the months prior to its initial public offering. This had led to speculation that the company was loosening its policy to increase revenue and ramp up its rate of growth prior to the public offering to get a higher stock price. Previously, Google, like Yahoo, had in general prevented companies from buying ads linked to trademarked names owned by others. If one slipped through and an advertiser complained, Google would pull the offending ad. Google now had a new approach and an explanation too: it didn't want to interfere in the free exchange of ideas that could benefit its users, who might want to see a range of competitive pricing and information. For example, typing the word "Geico" into a query box frequently produced not only ads from competitors but links to Web sites that offered price comparisons among a number of auto insurers.

In its 2004 public filing, Google said simply that it had changed its policies and practices "in order to provide users with more useful ads" and acknowledged that the change brought potential risks. "As a result of this change in policy, we may be subject to more trademark infringement lawsuits," it wrote. "Adverse results in these lawsuits may result in, or even compel, a change in this practice which could result in a loss of revenue for us."

Google already faced a number of other legal challenges on trademark issues. The most famous ongoing court fight was with a company called American Blind and Wallpaper Factory, which had been battling with the search engine over trademark issues that it claimed could put it out of business. None of the cases in the U.S. had been adjudicated, but Google had suffered legal setbacks in France, where it was trumped in court by luxury goods maker Louis Vuitton, and elsewhere in Europe, where the laws governing comparative advertising were more restrictive than in the U.S.

In certain respects, the case before Judge Brinkema was as old as advertising itself. Historically, the law allowed comparative, competitive advertising in the U.S. under a doctrine known as "fair use." This permitted advertisers to make claims about how their prices and products compared to named competitors, provided they did not make false claims. But in other, more profound, ways, Judge Brinkema had the potential to break fresh legal ground by taking on the case. Because the Internet was relatively new, as was the medium of search engines as advertising vehicles, no clearly developed body of case law or statutes existed for her to fall back on when rendering an opinion. In this respect, the case was on the legal frontier, the kind of place where Google found itself most comfortable as it created new domains for engagement in technology, business, culture, high finance, and, now, in the law itself.

For Geico, this was merely a fight about one particular advertising vehicle, a legal joust that the average company could not afford to bother taking on. For Google, it was a battle to protect its franchise.

As the *Geico* v. *Google* trial began, Judge Brinkema made it clear that she had read all the filings in the case, knew something about the businesses and the issues, and hoped to move things along at a crisp pace. Google had tried to avoid its day in court at a pretrial stage by filing a motion to have the case thrown out, but after reviewing the core issues, Brinkema denied the motion, ruling that there was sufficient evidence to let the trial proceed.

First up was Charles Ossola, the Arnold & Porter partner who took the lead in pressing Geico's case for trademark infringement. Geico had spent more than $1 billion promoting its brand name over the previous five years, and it was now being forced to either bid for a name it already owned or run the risk of having computer users type in "Geico" and then buy car insurance from a competitor whose name appeared in an ad. According to Geico studies, most people got a quote only from one auto insurance company and then bought insurance from that firm. This meant that competitors were winning customers at Geico's expense.

"Computer users are looking for Geico and an insurance quote, and instead they end up with a company that is not affiliated with Geico," Ossola said. "Geico wants a ruling that Google's sale of its trademark creates a likelihood of confusion."

Not so, countered Google's lead counsel, Michael Page, in his opening argument for the defense. Facing Judge Brinkema as he spoke, Page argued that Geico's counsel was making artificial distinctions between the Internet and traditional media.

"Geico wants to decide the Internet is different. That is based on an incorrect assumption that when people type 'Geico' into Google's search engine, the only thing they want is Geico's Web site. That is wrong." Page further argued that the legal doctrine of consumer confusion did not apply, noting that Google clearly separated the ads served up with search results, placing them to the right of Web site listings, on the other side of a blue vertical line. He also stated that Google, like a magazine or newspaper, was not the responsible party when it came to competitive advertising.

"Google is the publisher. The advertiser writes the text. Geico's argument might support a claim against an advertiser, but not against Google."

As is customary in such trials, both sides had expert witnesses, company officials and others ready to testify on their behalf. There would also be ample opportunity for the two sides to cross-examine witnesses. But on the third day, after thoroughly ripping apart Geico's lead expert witness and highlighting the weaknesses in a survey done by the company to bolster its case, Michael Page rose and made a motion to end the trial.

"At the opening of this case, I said that the Court would be able to dispose of this case before hearing testimony on damages. That got a bit of a laugh, but I was quite serious. Now that the Court has heard plaintiff's case, we believe you can rule in favor of Google as a matter of law," Page said. It was an integral part of Geico's business strategy, he explained, to sell car insurance direct to customers and not allow other companies to quote rates. "It's a perfectly legitimate strategy and, combined with a great marketing campaign, has resulted in a remarkable success story. That's their choice, but Google cannot be liable for confusion that's a natural part of their business strategy. What Geico evidence shows is that the use of trademarks as keywords does not itself create confusion. As they bear the burden of proof on this point, that's the end of the analysis."

While Page said Google's policy and systems were designed to prevent competitors from using the Geico name in the headline or text of their ads, he acknowledged that the system didn't work all the time and that improper ads sometimes slip through.

"Google vigorously enforces that policy. The inability to achieve perfect enforcement of that policy does not give rise to contributory liability." In order to prevail on that claim, he said, Geico would have to establish that Google encouraged or knowingly assisted in a violation of trademark law. He said there was no such evidence before the Court because Google did not condone or encourage infringement. "Quite to the contrary," he added, "we were the first search engine to implement a trademark enforcement policy, both because it is the right thing to do and because it makes business sense."

Page pointed out that Google served different sets of customers and had to be sensitive to multiple interests. "Trademark owners are our customers on the advertising side, just as the users are our customers on the search side," he said. "Thus, we carefully balance the dual goals of protecting trademark rights on the one hand, with providing our users with the most complete and relevant information possible on the other hand. We believe that our current trademark policy strikes exactly the correct balance, neither over- nor under-protecting trademark rights. Therefore, we ask that the Court enter judgment as a matter of law in Google's favor."

Having sucked the air right out of the courtroom, Page left Geico's lead lawyer, Charles Ossola, trying to find something of legal substance that might refute his rival barrister. Page seemed an embodiment of Google itself: entrepreneurial, spirited, and establishing superiority blow by blow. The old lawyers' adage still applied: when you have the facts, argue the facts; when you have the law, argue the law. Ossola rose to his feet to argue the law, as he was obliged to do, but he rambled, and if he emphasized anything at all, it was the facts, for those were what he had left in his arsenal.

Launching into a long, confounding description of the alleged wrongdoing that at times made Geico sound more like the defendant in the case than Google, Ossola argued: "We believe that the evidence clearly establishes that there is a likelihood of confusion based both upon the use of 'Geico' in the text, and in this case, given the fact that Geico is a [trademark] associated with rate quotes, that it is also sufficient to establish a likelihood of confusion with respect to those sponsored listings that do not contain 'Geico' in the text." And with that, he simply said, "Thank you, Your Honor," and sat down.

Peering down from her perch, Judge Brinkema wasted no time in ruling. "[Geico] has not established that the mere use of its trademark by Google as a search word or keyword standing alone causes confusion," she said, adding that there was insufficient evidence presented in Geico's case to let it go forward. "There's serious legal issues that are involved in this decision that I've just rendered, and I think I would like time to write a more detailed opinion on these legal issues. What I propose is that we terminate the trial at this point."

It was over. Four months after taking on Wall Street, Google had won a major legal victory against the Washington business and legal establishment. Predictably, the two companies would go on to issue dueling statements, with Google declaring victory and warning others not to challenge it on legal claims of trademark infringement. For Geico, it was back to business. For Google, the victory vindicated its ad policies and added to its enormous momentum.

Google still had to deal with its legal setbacks overseas, where the company was obligated by law to adhere to stricter policies on trademarks in ads. But it seemingly grew stronger and more confident through this period, having established the legal precedent in its most important market, the U.S., and having proved that it could adjust its practices elsewhere and still earn a hefty profit.

But just hours after Google's legal triumph in Virginia, something unrelated happened that took everyone's minds off the trial and its outcome: a breathtaking announcement that made the legal issues in the case seem small.

CHAPTER 21

A Virtual Library

The library of the University of Michigan, with some seven million volumes in its holdings, houses one of the largest collections of books in the United States and is an esteemed center for scholarly research that dates back more than a century and a half. For Larry Page, who had wandered among its stacks as an undergraduate, the Michigan library sparks a tinge of nostalgia for his not-so-long-ago college days. It also looms large in his plans for the future. He was on the hunt for untapped mountains of data that could be digitized and made searchable online. And Michigan was an Everest.

If the twenty-first century was truly to be an Information Age, Page and Brin knew the Internet had to expand beyond its current state. There was something to the joke that according to the Web, the world began in 1996. For all of the billions of documents in Google's database, from online encyclopedias to government documents to department store catalogs, all of them instantly searchable with a keyword and a click, combing the Web for serious, authoritative, time-tested information could still be a frustrating experience. The quality and reliability of information varied widely. As with television, surfers had to contend with channels and channels of junk as they looked for something of substance.

Page and Brin felt this frustration themselves, and after surveying the intellectual and technical challenges involved, the two of them devised a bold solution.

At a development dinner on Michigan's Ann Arbor campus in 2002, Page approached university officials with an audacious offer: he would foot the bill to scan every book in the Michigan library if he could add the information to Google's index. The idea was to put millions of books—the most important and most obvious repository of information—into a format where more people could access them. The Internet included plenty of information culled from books and about books, and even the complete texts of non-copyrighted tomes such as Shakespeare's works and the Bible. But the vast numbers of books published over the ages and in the present day were poorly represented.

Page and Brin were prepared to devote significant amounts of money and technological resources to digitizing the miles and miles of books that were gathering dust and growing brittle on rarely visited shelves in the great libraries of the world. The potential for the advancement of human knowledge was enormous. For the first time, scholars, and ordinary people too, would be able to access the full texts of books from a computer anywhere in the world. (For books still in copyright, users would be able to view only snippets of pages, although they could search the entire text.) With a trove of exclusive content, Google would have yet another way to distinguish itself from competitors, drive traffic to its site, and profit from relevant ads that could be displayed alongside the book information. Moreover, if it became adept at book digitization, other libraries would be more likely to sign on, creating an even larger base of searchable content.

Page and Brin faced significant hurdles. They would need to perfect a technology that enabled them to digitize efficiently and accurately on a scale never before attempted. They also had to find libraries willing to go along with the idea, which for a privately held company only a few years old would take some convincing. This was one reason for starting at Michigan, Page's alma mater. Because he was a notable alumnus and still closely involved with the school, he was likely to get the opportunity to make his case fully.

A few months after the Ann Arbor dinner, Page sat down with two of the university's librarians, William Gosling and John Wilkin. They were receptive to Page's proposal from the start, in part because book digitization was on their own wish list and they had yet to find either the funding or a suitable partner for the project. In prior discussions with IBM and other corporations about digitizing some of Michigan's collections, Wilkin said, "We were so far apart in terms of intellectual property, intentions in doing the work, the rights to the works, standards, formats, all these things." But with Page, there was a level of confidence and trust from the start. "When we sat down with Larry, we came away feeling something distinctly different."

Around the same time, Page was also floating the idea at Stanford, the other university where he maintained close ties. The school had been the fountainhead for Page's idea for bringing offline content onto the Internet. He had dreamed of it when working on the Digital Libraries Project with Professor Terry Winograd. Though that project hadn't been specifically about digitizing books, Page realized then that everything could be saved in digital form and displayed online, and he began saying, "We'll scan it all.

"Even before we started Google, we dreamed of making the incredible breadth of information that librarians so lovingly organize searchable online," Page recalled. But as with other Google innovations, including the original search engine itself, a major reason Larry and Sergey pursued the massive scanning of books was simply to satisfy their own curiosity.

"I just wanted to be able to search libraries myself," Larry said. "You get interested in something and want to see the state of human knowledge." He believed the digitizing and posting of books online would have a big impact on scholars by freeing them from the need to travel to many different libraries to do research. "Right now, it is really hard for scholars to work outside of their area of expertise because of the physical limitations of libraries." He was reminded of his frustrations as a kid in Michigan when he couldn't get his hands on the manuals he needed for reassembling electronics gadgets that he had dismantled. "There was this one bookstore where you could get them—only one." The information inside those manuals was only accessible in that single location. It was much the same with aging books tucked away in university libraries. It didn't matter if there was a ready audience for the information; there was simply no mechanism for distributing it to the masses.

Page brought his vision to a private conference held by Paul Allen, the Microsoft cofounder and billionaire philanthropist, at his retreat in the San Juan Islands in Puget Sound, near Seattle. Allen had invited a small group of technologists and academics to brainstorm a project of his called the "Final Encyclopedia," a broad look at assembling all the information in the world. Stanford's Terry Winograd arranged for Page to attend in his place, and he used the occasion to raise the idea of a mass digitization project with Stanford Library director Michael Keller, stressing the importance of adding books to the searchable universe of the Web. Keller concurred, and earnest discussion with the university began.

"The idea of large-scale digitization is something Stanford was interested in on its own," said Stanford librarian Andrew Herkovic. "Google presented us with a lucky break on something we've been trying to figure out how to do for many years."

Sidney Verba, the director of the Harvard University Library, had heard all the various schemes out there for scanning books into digital form, even one that involved boxing and shipping rare volumes overseas where the labor-intensive work could be done more cheaply. This was out of the question for the hallowed institution Verba was charged with overseeing. Harvard's library, founded in 1638, is the largest academic library system in the world. Verba and the other library trustees had discussed the options for digitizing the massive Harvard collection, some 15 million volumes, but concluded that the schemes they had seen would be incredibly expensive and would damage the books.

So it was with some skepticism that Verba received a representative of Google in 2002 who had come to ask if he would be interested in having the entire Harvard library digitized. "I don't know how to put this—I had the feeling they were smoking something," Verba recalled, though he described the initial conversation as cordial. They touched on questions of cost, disruption to the collection, and possible damage to books. "I thought it would be terrific if they could do it," Verba said. "But I expressed at least as much skepticism as optimism. I thought it would be a long time before anyone digitized a whole library."

Several months later, a team of Google's top product development specialists returned to Harvard. Their updated presentation described "a much more efficient, less damaging means of scanning books," Verba said. "I never thought it would come off when I first heard about it, but when they came back, they had done a tremendous amount of work on the technical side and thinking about the issues. It looked quite real." So he and a team of Harvard librarians traveled to the Googleplex to see the new scanning devices for themselves. "We were impressed," Verba said. "What was really impressive was that they had gone and engineered a new mechanical means of scanning that does the work of putting books into a digital form. Something that was quite efficient and treated the books much more gently than other scanning devices people had been marketing. I think all of us were turned."

The next step was to bring the prospect of the project before the Harvard Corporation, the university's governing body. This was a group for whom caution ruled the day. The corporation wanted reassurances that the university wouldn't be hit with lawsuits for violating copyrights, or incur heavy costs, even with Google's offer to pick up the tab for the scanning project. "If someone does something for you, it typically costs you money as well," Verba said. Then, too, there would be an old, esteemed university partnering with an upstart technology company in the wake of the dot-com crash. Google had only been around for six years, but Harvard was 375 years old.

By the late 1990s, the technology for scanning printed materials into digital form had become commonplace: many offices had some type of flatbed scanner for converting documents, photographs, or other art into computer files. But the process was not readily scalable to the level of millions of books that Google hoped to scan. The company needed to develop a means to digitize the books at high speed, and at a high level of accuracy, if it was to make the project worth doing.

In the spring of 2003, teams from Google and Michigan met to review scanning methods and technology. In the past, Michigan had not shied away from dis-binding the books it digitized—literally removing the binding to be able to scan them flatter and faster. This sped up the process, but many librarians frowned upon the practice because it damaged the original books. Google too had experimented with so-called "destructive" scanning methods: for its Froogle shopping service and an on-again, off-again project called Google Catalogs, it chopped the backs off of retail catalogs and shot individual pages through a sheet feeder that resembled a high-performance photocopier. Given the large scale of Google's proposed digitization, however, both sides agreed that dis-binding was not an option; too many books would be affected. So the Google team set out to design a system that moved at lightspeed but had a light touch. They looked at automated devices that turned pages by robotic arms or by air suction, eliminating the human handling of books, which can be damaging even with innocent movements. And in the end, they opted for a process that involved more manual labor, adopting others' technology to suit their particular needs. They would hire a small army of personnel who would be carefully trained to handle the books and scanning machines.

"We didn't build the camera. We didn't build the scanning," said Adam Smith, a project manager on the library effort. "It's been as much about process and system as about technology."

Wilkin was impressed with Google's solution. "It took them longer than they thought," he said, "but they did it."

Reg Carr, the affable director of Oxford University's Bodleian Library, was in the midst of raising money for digitization plans of his own in 2002, when he struck up a conversation with Raymond Nasr, Google's director of executive communications. Nasr was at Oxford speaking to a group of business students as part of a "Silicon Valley at Oxford" program, and Carr took the opportunity to show him around the storied gothic campus, letting its rich history work its own magic on his guest. Unaware of the developments at Michigan and elsewhere, Carr raised the possibility of a collaboration between Google and the Bodleian.

"My pitch was: Bodleian in terms of libraries was first in class, and Google in terms of information technology and search is first in class, so shouldn't we be doing something in partnership?" said Carr. In his world of fund-raising, a prime activity is finding "how to get resources for something you already want to do." Because the Bodleian Library at Oxford is a depository library, meaning it receives a copy of every book printed in the United Kingdom, just as the Library of Congress does in the United States, Carr believed it had special appeal for Google. It was also the second largest library in Britain.

If there was ever any doubt about Google's interest in adding Oxford to the list of libraries, it was put to rest in the fall of 2003 as Carr was planning a fund-raising dinner in San Francisco. Nasr joined the dinner-organizing committee and helped to sell tables, of which Google bought several. Carr used the opportunity to solidify the relationship and show off a bit. He rented out the elegant Butterfield's in downtown San Francisco for the black-tie affair. "Google people were there in force," he said. To add the right touch, he had arranged to have the banquet room kitted out in the style of an Oxford college. Armed guards stood watch over 50 million British pounds' worth of rare books transported to California for the event. Inside, Larry, Sergey, and other Googlers chatted and had drinks around cases with treasured works, including a copy of the Magna Carta and an original illuminated manuscript of Chaucer's *Canterbury Tales.*

Despite Google's presence at the Oxford dinner, the entire library digitization project was still very much under wraps. Google insisted on absolute secrecy about all aspects of the library project, which put some of its university partners in an awkward situation. "People seem to be in mortal terror of the nondisclosure agreement that Google imposed," Stanford's Herkovic said. Google had required everyone with knowledge of the project to sign a very strict nondisclosure agreement (NDA), typical in sensitive business dealings but almost unheard of in the library community.

Paul LeClerc, head of the New York Public Library, who was also in discussions with Google to digitize a portion of its vast research collection, said, "I don't know if I've ever signed an NDA before." Harvard's Verba understood their reasons for keeping mum, but he was uncomfortable. "Google was not forthcoming about with whom they were talking; they prefer to operate in a quiet way. Universities are basically open institutions. There was something awkward about not being able to tell the faculty that this was coming down the pike." For a community based around the open dissemination of information, it was peculiar, even paradoxical, to be sworn to secrecy.

On February 1, 2004, the secret almost got out. A *New York Times* article on Google's prospects for the future had a fleeting mention of a book digitization effort with Stanford code-named "Project Ocean." The library community buzzed about what might be coming, but no one was able to shake loose any more information.

In July, as Google furiously prepared for its IPO, the Google Print team moved the first scanning devices to the University of Michigan, and began training the contractors who would operate them. Michigan was to be the test ground for the new devices and the logistics of scanning. When the truck arrived with the scanning equipment, a person not involved with the project began asking questions about what was going on. Wilkin realized they needed a strategy on how to preserve the secrecy, and fast. The solution they arrived at was to hide it in plain view. They set up in an area of the library where other scanning was done, and on the door of the room with the special Google equipment they posted a sign that read MICHIGAN DIGITIZATION PROJECT.

"And that was the last we heard anything," Wilkin said.

The technology for scanning was not the only challenging part of getting the project off the ground. Google had aggressively addressed most of the concerns each library had about physical care of its books. That was important to librarians and archivists looking to preserve the collection for posterity. But what about the part of a book that the university didn't own—the copyright? This was the crux of what Google was after: the sentences on the page, the ideas and text that Google could index and search—what lawyers call "intellectual property." For nearly every book published in the United States after 1923, ownership of the copyright resided with the publisher, the author, or some third party.

Copyright law in the U.S. is a tricky field, full of conflicting interests and a great deal of misunderstanding. Google dove right into that briar patch. Copyright protection enables publishing companies and authors to make money from creative works. Though the library project at its core was about the free dissemination of knowledge, a lot of money was at stake. And that meant it was probably only a matter of time before groups protested and lodged legal challenges. "It would be naïve to assume that this activity will not be tested in court," Herkovic said. "A lot of our hopes and ambitions will be determined by how things come out in the wash after some amount of litigation. Google has their eyes open in this. They know they'll get nailed—or hit on. We, Stanford, will presumably not be able to just watch from the sidelines." To address this issue, Google agreed to reimburse the various libraries for any costs arising from lawsuits over the digitization of their books.

Mike Keller, the Stanford librarian, believed Google's approach was smart. "By gathering five very large libraries together, we have the opportunity to work on some of the more thorny issues—the copyright regimes in the U.S. and U.K.," he said.

"Getting the publishers' support is important," Page acknowledged. "For this to work, we have to make sure publishers make money." He recognized that Google's assurances might sound like lip service. Some way to demonstrate the company's commitment would go a lot further.

In early October 2004, Larry and Sergey attended the Frankfurt Book Fair in Germany, the global publishing industry's most important annual meeting, to unveil their latest search tool: Google Print. With typical Google gusto, including flashing lapel pins and free T-shirts, Brin and Page, dressed in suits and ties, sat for a 40-minute press conference, during which they explained the details of how they would make copyrighted books accessible and searchable online, presumably without legal action or fuss from publishers. Their idea was not entirely new—Amazon.com had launched something similar a year earlier with its "Search Inside This Book"—but its release was carefully timed.

The publishing industry wasn't quite sure whether to love or hate Google. On the one hand, they shared its interest in disseminating information and could profit from additional exposure. On the other hand, Google sought to distribute content as widely as possible for free and make its money from advertising, while publishers hoped to sell books. In the U.S. alone, over 1,000 books were published each week. That was a staggering amount of content, and Google, sizing up this potential for information and profit, offered publishers a deal of sorts. It would cover the costs of scanning and indexing books for the right to display them as part of search results. It would display only the few selected pages or snippets of text that related to a user's query, and in a form that presumably couldn't be copied or printed. Google would provide bibliographical information and direct links to booksellers, and would share with the book's publisher the revenue from any ads it displayed.

Google's idea was simple: giving the reader a taste of the book to entice him into purchasing a copy. By the time of the Frankfurt convention, most of the major publishers had signed on to Google Print. (Amazon's program already had some 120,000 titles and 33 million pages of text available.) Perhaps more important, it won support from publishers for the idea of displaying copyrighted content on Google before the supersecret library project was unveiled.

On December 14, 2004, Google revealed at long last its intention to digitize 15 million library books. *The New York Times*, the old grey lady of print, carried the news as its lead story on the front page. "This is a great leap forward," said Michael Keller; it would take Stanford's digitization to a "truly industrial" scale. Some likened the potential significance to that of Gutenberg's printing press. "The world changed today," Michigan's Wilkin said.

Even at the start, the scale was daunting. Google's agreement called for digitization of all seven million volumes at Michigan; the million or more volumes in Oxford's nineteenth-century collection; 40,000 books from Harvard; 12,000 from the New York Public Library; and an unspecified number from Stanford. If the libraries that signed on for only a modest pilot program agreed to go ahead with their entire collection, Google would have more than 50 million complete books in its database when the scanning was finished, perhaps in a decade. By comparison, Page and Brin had launched Google in 1998 with an index of 25 million Web *pages*.

Not everyone viewed the digitization project as a good thing. After the announcement, groups representing publishers, authors, and librarians, which felt threatened by Google's plans, attempted to check the exuberance and derail the project. In an op-ed in the *Los Angeles Times,* Michael Gorman, then president-elect of the American Library Association, downplayed the efforts of the "boogie-woogie Google boys." Massive databases of digitized books are "expensive exercises in futility," he wrote, "based on the staggering notion that, for the first time in history, one form of communication (electronic) will supplant and obliterate all previous forms. This latest version of Google hype will no doubt join taking personal commuter helicopters to work and carrying the Library of Congress in a briefcase on microfilm as 'back to the future' failures, for the simple reason that they were solutions in search of a problem."

In France, the project was looked upon as yet another example of America flexing its cultural muscle. Jean-Noël Jeanneney, head of the venerable Bibliothèque Nationale, wrote in the newspaper *Le Monde* that the Google venture posed "a risk of crushing domination by America in defining the idea that future generations have of the world." Jeanneney said, "I don't want the French Revolution retold just by books chosen by the United States."

CHAPTER 22
Trick Clicks

Samuel Baruki Cohen, the 34-year-old president and founder of lendingexpert.com, sensed that something had gone awry on Google. A former executive at Merrill Lynch and Bloomberg Financial Markets, he had become a successful entrepreneur by applying business lessons that he learned from New York City mayor Michael Bloomberg, the founder of Bloomberg Financial. Among other things, Cohen had learned to spring into action aggressively and immediately if something seemed questionable. In early 2005, that is exactly what he did.

Until then, Cohen had been satisfied with his relationship with Google, which had enabled him to build a profitable online business dealing in home mortgages. At lendingexpert.com, the third business Cohen had started since 1997, he employed a simple strategy: spend heavily on marketing to ensure that ads for his firm appeared prominently alongside Google search results. This Internet-only marketing approach held down costs, and at the same time, by associating his firm with Google, gave lendingexpert.com the kind of instant credibility it needed to attract thousands of customers daily and compete effectively with its better known rival, lendingtree.com. The business model was simple too. When computer users clicked on one of lendingexpert.com's ads, they were invited to fill out a preliminary mortgage application.

The firm then profited by selling the applications, which were really business leads, to mortgage brokers and others, who paid handsome fees to gain access to new customers hunting for home mortgages.

The trouble began when Cohen noticed that an increasing number of clicks on lendingexpert.com's ads looked suspicious. For one thing, many of the clicks came from the same Internet addresses. And none of the clicks was generating any business. In other words, someone, or some business, or some automated software, was clicking away on lendingexpert's ads, driving up its advertising costs without ever completing a mortgage application. Cohen knew roughly what proportion of users who clicked on his Google ads went on to fill out mortgage applications, making the barrage of dubious clicks easy to spot. The unanswered questions were: how much was this costing Cohen in advertising fees his firm paid to Google, how could it be stopped, who was behind the artificial clicks, and what was motivating them to hurt his business in this way. Though he couldn't prove it, Cohen sensed that a competitor was clicking on his ads and trying to damage his business by driving up his costs.

"Our whole business is bringing people to our Web site to fill out mortgage applications and then selling those applications to mortgage companies," Cohen said. "Our biggest way of doing that is through Google and other search engines. We advertise heavily on Google, which brings us the most clicks. What happens in a competitive industry, like mortgages or insurance, is that your competitors know you are paying Google per click. If somebody is out there and they get somebody to click on your pay-per-click ad a thousand times each day, and you are paying $5 to $10 per click, it is a lot of money."

Cohen couldn't easily pull back on advertising or switch to traditional ads, given his business model. "We are very vulnerable," he said. "We need to be out there selling to mortgage companies, and they want to know we have a major presence. If you go onto Google and type in 'commercial loans' or 'construction loans,' we need to be there. We need to be on search engines to stay in business and compete and succeed. People using search engines are more likely to fill out an application for a mortgage than somebody reading a newspaper or reading another news source on the Net.

"What happened on Google is that we suspected some click fraud," Cohen said. "We average a 5 percent conversion rate—for every 100 people who click on our ads, we expect five to fill out applications. We were getting an excessive number of clicks without any applications and started looking into it. In one day, one IP address hit us 30 or 40 times. It was excessive."

To handle the matter, Cohen turned to Nicole Berg, who runs his firm's search engine marketing program and closely tracks its performance, morning, noon, and night. After compiling the basic data, Berg, who logs on numerous times daily from both work and home to monitor activity, agreed there was a problem. Berg sent an email about it to Google on January 6, 2005.

> We are the victim of click fraud. Our systems administrator was able to track down the IP address of someone who has clicked on our ad 20 times in one day which has cost us $200. We would like to request a refund.
>
> What information do you need to refund us?
>
> Thank you.

"We didn't expect them to respond primarily because I was assuming it was very prevalent and difficult to really track it down and stop it," said Cohen. "I was not expecting much. On any given day, we might get 2,000 clicks on Google. . . . Who is going to look through all those clicks and determine where they are coming from? How you can do that kind of detection on such a large scale I'm not sure."

Five days later, Berg received an automated reply from Google.

> Hello Nicole,
>
> Thank you for submitting the report of suspicious activity on your AdWords account.

If you are able to supply any of the following information related to the suspicious activity, it may help us significantly expedite our investigation:

- The Campaign(s), Ad Group(s), and/or keyword(s) associated with the suspicious clicks.
- The date(s) and time(s) of the suspicious click activity.
- A paragraph describing the trends that led you to find the click activity suspicious.
- If you have access to Weblogs or reports, any data indicating suspicious IP addresses, referrers, or requests.

Upon receiving this information, we will work as quickly as possible to investigate your issue.

Please feel free to reply to this email if you have additional questions or concerns.

Sincerely,
The Google Click Quality Team.

Click. Click. Click. To Google and its founders Sergey Brin and Larry Page, that was the sweet sound of money, the Google equivalent of a cyber–cash register ringing up sales, as millions of computer users around the world clicked on the small text ads that appeared above and to the right of Google's free search results. Each click of the mouse brought Google anywhere from a nickel to more than $50 from advertisers, who paid by credit card for the chance to have their products or services described in small, square boxes alongside Google search results. With computer users clicking at all times of day and night in every corner of the world, from work, home, school, and Internet cafés, money poured into Google in dozens of currencies.

Google brought in more than $1 billion annually from click-based advertising—an average of more than $100 million a month, or more than $3 million a day every day of the year, whether its employees were awake, asleep, playing beach volleyball, or on a company-sponsored ski trip. With funds streaming in nonstop via the Internet, Brin and Page knew it was just a matter of time before some would employ unscrupulous means to disrupt this commerce or attempt to reach in and grab some of that honey for themselves.

Google chief financial officer George Reyes has described click fraud as the most significant threat facing Google's highly successful business model and long-term profitability. Click fraud often manifested itself in one of two ways: businesses clicking away on the text ads of competitors in order to raise their marketing costs, or Web site publishers who were part of Google's affiliate network repeatedly clicking on ads served to their own sites in order to pocket more revenue. Whatever its form, click fraud involved automated or manual clicks that had no chance of producing business leads, costing advertisers money and exposing a vulnerability in Google's cost-per-click advertising model.

There were ways to shut down click fraud, but virtually all of them would make Google less profitable. It could employ armies of people to work on the problem, and it has employed some, but not as many as it had working on new products. It put filters on its ad system to block some very obvious forms of click fraud, but not so many that potentially profitable clicks would be blocked. One example of a remedy Google has not tried is that of Snap.com, a search engine that used a cost-per-action method of billing, where advertisers were charged only when computer users both clicked on their ads and completed a specified transaction, whether it was making a purchase or filling out an application. Google charged advertisers for all clicks, whether or not they led to purchases. Snap.com grew out of the same incubator on the West Coast, Idealab, that had earlier given birth to cost-per-click advertising, and its new approach, though less profitable, may be the wave of the future if competition heightens in the search engine space, or if click fraud becomes such a pervasive problem that Google and Yahoo are forced by advertisers to make concessions.

Tom McGovern, the chief executive officer of Snap, believes his company's approach gives advertisers greater protection from click fraud, especially those who do not monitor online performance closely. "Online advertisers are being 'clicked and dimed' to death" under the cost-per-click model, McGovern said. "Snap is the first paid search vehicle where the interests of the search engine and advertisers are aligned." Since advertisers paid only when a transaction occurred, this increased their financial return on marketing dollars spent, he said. Instead of paying per click the Google and Yahoo way, advertisers on Snap could choose to pay for various types of business results—sales, downloads, leads, subscriptions, or other actions important to the marketer. This next-generation search engine system, according to Snap, guaranteed that every marketing dollar would yield a customer.

Lacking cyberpolice to hunt down click-fraud perpetrators, the Google Economy has responded mostly with free-market solutions—self-policing by Google and Yahoo, and the creation of a new industry employing hundreds of people who specialize in fighting this flavor of fraud. But this is an industry problem being addressed on a piecemeal basis. No search engine—not Google or Yahoo or any other—has the financial incentive to put sufficient resources into battling click fraud, since the search engines profit from its existence and it mostly goes undetected and unreported by advertisers. Though individual advertisers can amass plenty of information to illustrate that they appear to be the victims of click fraud, while acting alone, they lack the means to stop it.

The private firms that battle click fraud on behalf of advertisers seek to persuade Google and other search engines that they have compiled the data that proves it occurred and demand refunds. In relative terms, advertisers on Google have suffered most from the problem, largely but not exclusively because Google is the biggest player in this rapidly growing marketing segment. If not combated, click fraud may undermine confidence so much that a self-correcting mechanism takes hold. Advertisers may cut back on how much they are willing to spend, or demand a new system outright. Another possibility is that major search engines will join forces to fight the problem as an industry, as major Internet services did to combat email spam.

"There is going to be click fraud," said Andy Beal, a search industry analyst. "But paid search works well and is an extremely effective method to get business to a Web site. Over a period of time, you might see advertisers getting a 4.5 percent conversion rate instead of 5 percent. Instead of being willing to spend $2 per click, they may bid $1.90 per click. People will take it for granted that there will be some clicks of a suspicious nature but the model will be robust enough to withstand that. I don't think it is going to be an epidemic. People will take it into account and price it in."

However, Robert Deignan, vice president of Internet security firm STOPzilla, considers click fraud a serious problem that must be addressed. "It is frightening to realize all of these clicks are streaming in that are not real traffic," he says. "It is very hard to prove, but it *is* fraud. At the end of the day, it is fraud. It is very hard to police. It is sad. Groups are spending fortunes on what they think is legitimate traffic and it is not. Ninety-nine percent of the time we don't have a problem getting a refund from Google or Yahoo. If there is a problem, they often find out before we do and say we just credited your account. If there is a problem we do see first, it is refunded in a day or so. They are solid. But a lot of advertisers don't have the tools to realize what is happening."

Jessie Stricchiola fell into the business of fighting click fraud by chance several years ago, when she was handling online marketing for the Chase Law Group. Since that time, she has parlayed the opportunity into a passion and a specialty. As founder and president of Alchemist Media, Inc., she spends her time waging battles with Google and Yahoo on behalf of advertisers seeking refunds, or providing advertisers with the data they need to challenge the major search engines themselves. In the course of her work, Stricchiola has encountered many different varieties of click fraud.

She vividly recalls the story of a consumer electronics firm that advertised on Google and Yahoo and suddenly had a sharp spike in the number of clicks on its ads that were not converting into new business. The firm identified the source of the automated clicks as an Internet address affiliated with one of its primary competitors.

The problems, which had begun in 2003, continued despite a six-month effort by the electronics firm to persuade Google and Yahoo that this was click fraud. Having failed to win relief, the firm hired Alchemist Media for help. Then there was a strange twist. The firm received an anonymous email: "I just want you to know that [a competitor] is intentionally clicking on your ads and costing you a lot of money. I know because I used to work for them and developed the software." Attached to the email was a video depicting how the click fraud was being carried out without human intervention. The software was sophisticated enough to avoid repeating patterns, and each click was costing the advertiser anywhere from $6 to $15.

After Stricchiola became involved, Yahoo refunded money to her client, but Google declined. Google's stance left the advertiser, who is now pursuing a lawsuit against its competitor, with hundreds of thousands of dollars of losses from advertising expenses. Stricchiola was convinced it was garden-variety click fraud. She added that numerous other click-fraud cases of advertisers alleging wrongdoing on Google or Yahoo have been settled out of court and remain under seal. "I am aware of a number of those," she said. In general, she considers Yahoo more responsive than Google to advertiser complaints about click fraud.

"Google is notorious for just flat-out ignoring advertisers," she said. "Google says, 'Thank you for your inquiry. We see no problem.' Sometimes Google does not even look at the data, and they give the most ridiculous explanations. Yahoo tends to be more proactive. That is because they have the people, more so than Google, to really look at this stuff. This is why some advertisers tear their hair out and I get hired to help people and create a case at what is looked at as a purely technological issue at Google." She added that Yahoo shares data about click fraud with advertisers, while Google, which sometimes sends advertisers refunds with no detailed explanation, does not, arguing that it does not want to release information about detection methods that could fall into the wrong hands and lead to foul play.

Google officials disagree with Stricchiola's description of their approach to addressing click fraud. "We believe we can manage click spam very effectively," said Google product manager Salar Kamangar, using the term *spam* rather than *fraud* to make the point that not all questionable clicks are malicious in intent.

"I would characterize the losses due to click fraud as small," he said. "We have a software system that filters out fraudulent clicks even before advertisers get billed for them. We are conservative with what we count, and throw out anything that looks suspicious. We also have a team of engineers and are constantly looking for ways to update the software. We also have a team of specialists investigating reports from customers that contact us and let us know they think there is a problem. We have prosecuted a case of this, but this becomes more difficult across borders. So our main focus is on catching it in the first place, detecting it, and giving advertisers the tools they need to track the effectiveness of their spending and the possibility of any click fraud."

In the case of lendingexpert.com, Google's timely reply, and subsequent refund, revealed several realties of click-based advertising and its risks. First, unlike many of the smaller search engines, Google responds to many advertisers who claim they are due refunds from alleged click fraud and has a department dedicated to working on specific cases and the overall problem. Second, the burden of proof of alleged click fraud rests with the advertiser, not with the search engine. Third, unlike the way credit card companies respond to customers—taking them at their word and preventing them from being billed for allegedly fraudulent charges while an investigation is taking place—Google is in a more powerful position, enabling it to collect revenue on the front end and decide later whether to offer refunds to an advertiser that makes a claim.

In other words, Google has the data, but not the incentive, to put sufficient resources into fighting click fraud, a growing problem. Conversely, many advertisers have the financial motivation to recover losses from click fraud but lack the comprehensive information they need to support their claims. A recent industry study estimated that as many as 30 percent of clicks may be fraudulent. It said that while Google is good at fighting click fraud that emanates from a business or person repeatedly clicking on an ad from a single Internet address, it is less adept at detecting click fraud that employs sophisticated software to mask its identity.

Some sites that sign up for Google's affiliate program turn out to be nothing more than fronts or shams set up to generate ad revenue. To deter such behavior and to highlight its vigilance in opposing click fraud, Google filed a $50,000 lawsuit in 2004 against a sham Web site called AuctionsExpert.com. To some, the lawsuit looked more like window dressing, a tactic intended to send a message by making an example of a Web site, rather than a major effort by Google to expend substantial resources to combat the problem.

Yahoo, for its part, has certain controls to hold down the level of click fraud that Google does not. For example, new Web sites that want to display Google ads can sign up online in a matter of minutes, whereas Yahoo reviews each new site manually, with people involved. Google apparently lacks sufficient controls to prevent shady Web sites from signing up merely to generate ad revenue through self-clicking. It appears more interested in recruiting new partner sites around the world for branding purposes, analysts say. "Click fraud is an issue we take very seriously," CFO George Reyes said. "While we cannot promise that we will be able to prevent all instances of click fraud, we are very pleased with the results to date."

CHAPTER 23
Attacking Microsoft

The room was packed with hundreds of eager computer science students and their professors by the time Eric Schmidt arrived at the University of Washington on a spring afternoon in May of 2005. With his jacket off, tie loosened, and BlackBerry strapped to his belt, the CEO of the world's most dynamic technology company moved briskly down the steps to the front of the lecture hall, primed to minimize Microsoft on its home turf and pump up Google. While many industry experts compared the two companies by focusing on software strategies and products, they tended to overlook the primary battleground. This was it. Given their relative strengths in differing businesses and technologies, the struggle between Google and Microsoft was not primarily about market share, browsers, computer operating systems, or whether only one of the two companies ultimately would survive. The real battle being waged was over recruiting and retaining the brightest technologists in the world. This was the key variable that would determine which company had the greatest capacity to identify and solve the most interesting and important new problems in the Internet Age.

Some months earlier, Google had brazenly opened an office in Kirkland, Washington, not far from Microsoft headquarters, where it had succeeded in luring away an array of talented software engineers who wanted to work for Google but continue to live in Seattle. Now Schmidt found himself behind enemy lines, inside the University of Washington's Paul G. Allen Center for Computer Science & Engineering—a building named and paid for by Microsoft cofounder Paul Allen. Schmidt was on a mission. He was determined to persuade the faculty and students at the highly regarded school that Google was a better, more exciting place to work these days than Microsoft.

Before he took center stage, Schmidt was warmly introduced by Professor Ed Lazowska, who held the Bill & Melinda Gates Chair in Computer Science & Engineering at the university, and whose motto was, "If you're not part of the steamroller, you're part of the road." Brimming with confidence, Schmidt was flying high and Google was on a roll: its stock price was topping $250 a share and the company's value exceeded $70 billion. The main question, from his perspective, seemed to be how many of the students seated before him would want to join Google on its historic journey.

For months—and even earlier that same day in another venue—when asked about competing with Microsoft, Schmidt had been both deferential and dismissive, saying Microsoft was a much bigger corporation with deeper pockets and greater resources, and that any comparison with Google was silly. But here his message had an edge: Google was the best place on earth to work, and Microsoft was an aging giant whose best days were behind it. To hear the world according to Schmidt, the company led by Gates was rooted in an earlier phase of technology that had been surpassed by the power of the Internet revolution. From this perspective, Microsoft mattered less and less, and Google mattered more and more. "You are playing on a much bigger stage with much bigger stakes," he said.

Then Schmidt drove in the dagger.

"We compete with Yahoo every day," he said. "Microsoft has announced their entry into the market, though they are not a significant competitor yet, although I am sure they will try." Twisting the knife further, Schmidt added, "Microsoft has publicly announced that they are going to imbed search in every single bit there is on a PC screen. So that is going to be their entry, but that will be some months, or years, depending on what schedule you assume."

It had become abundantly clear to Schmidt that Gates was reeling from the pressure of Google's extraordinary buzz, momentum, and rising stock price when the two men met at a technology conference earlier that same week. Only months before, people had been talking about whether Google would become the next Netscape—a firm that symbolized the Internet boom of the late 1990s before being crushed by Microsoft. Now people were talking about whether Google was overtaking Microsoft as the global leader in software, technology, and innovation. Gates, the richest man in America, constantly found himself on the defensive as he responded to tough questions about the competitive threat posed by the Silicon Valley search engine firm.

"Google is still, you know, perfect," Gates had sarcastically told the journalists and technology experts gathered at the conference. "The bubble is still floating. You should buy their stock at any price. We had a ten-year period like that."

It wasn't lost on Gates or any other savvy technology and investment mavens that from 1998 to 2005, Microsoft's stock was essentially flat, while Google had gone from being a start-up to a juggernaut. It pained Gates that Microsoft had been talking about making major strides in Internet search for years, yet its initial "MSN Search" offering several months earlier had not impressed most people. Still, with a Microsoft operating system controlling the personal computer desktop, Gates was determined to try to pick up millions of users by building a search engine into Vista, the next version of the Windows operating system. He would build an easy-to-use search engine that would be thrust in front of users whenever they booted up their personal computers. That was the plan.

"If anything touches on search," Gates promised, "we're going to do it."

While Microsoft was sprawling and deep-pocketed, it had lagged in search innovation, in part because the company had an abundance of software gurus but a dearth of expertise in the specialized field of search. Schmidt knew that budding software engineers would be impressed with a chance to go to work at a place like Google, where as part of small teams working on high-visibility, high-impact projects, they could make a difference on a global scale. For an ambitious geek or technologist, there was no greater motivation.

The pummeling Schmidt gave Microsoft that day was the clearest sign yet of just how ambitious and cocksure Google had become: nothing Microsoft could do would get in its way or disrupt its growth. "Our users," Schmidt said, "are everywhere." On the projection screen above him he flashed a slide of a mock *Newsweek* cover, showing what might have been if the Google Guys had gone to work for Gates: a photo of Brin and Page over a giant headline that read, "The New Age of Microsoft: Two New Guys at MSN are Making Redmond the King of Search."

Schmidt's march through Seattle that May took place against the backdrop of a devastating, high-profile article that had appeared earlier that month in *Fortune* magazine, which, along with *The Wall Street Journal*, was required reading for the American business establishment. The piece, written by *Fortune* reporter Fred Vogelstein, carried the headline: "Why Google Scares Bill Gates." It opened with the scene of a moment in late 2003 when Gates realized just how vulnerable Microsoft could be. The article portrayed a dire and deeply personal competitive scenario, coining the cunning catchphrase "Gates vs. Google."

Unlike other Microsoft competitors that had come along, Google, with its Internet-based architecture and penchant for developing free software with no cost of distribution and marketing, had attracted millions of computer users around the world. Microsoft had fended off other competitors with its suite of integrated office products more easily than it would be able dismiss Google, which was building a similar set of complementary tools using its own in-house software expertise. Already, Google had embarrassed Microsoft by beating it to market with the free desktop search product that helped computer users find information on their own Windows PCs. To make matters worse, Microsoft's Internet Explorer browser had been the starting place for Web search for years; now millions of people were turning to an upstart called Firefox, which had funding from Google and had Google search built-in. Microsoft's control of the computer desktop clearly would be loosened if Internet Explorer was in jeopardy as a browser, because its dominant products, from Microsoft Word to the Excel spreadsheet, were all programmed to activate and operate fluidly in conjunction with the browser.

"Microsoft was already months into a massive project aimed at taking down Google when the truth began to dawn on Bill Gates," the *Fortune* article said. "It was December 2003. He was poking around on the Google company Web site and came across a help-wanted page with descriptions of all the job openings at Google. Why, he wondered, were the qualifications for so many of them identical to Microsoft job specs?" The search company was posting job notices for engineers trained in operating system design, distributed-systems architecture, and other fields that were more Microsoft's provenance than Google's. "Gates wondered whether Microsoft might be facing much more than a war in search," the article said. "An email he sent to a handful of execs that day said, 'We have to watch these guys. It looks like they are building something to compete with us.'"

The "something" wasn't altogether clear, since the litany of new products that Google had been unveiling—from photo-sorting software to Web-based email to search via mobile phone—did not add up to a grand unified assault on the Microsoft desktop. But it did make users who adopted these services much less dependent on Gates's software, and potentially gave users a way to write, post, share, and print documents without using Windows and Word—the bedrock of Microsoft's empire.

"Google is interesting not just because of Web search, but because they're going to try to take that and use it to get into other parts of software," Gates told *Fortune*. "If all there was was search, you really shouldn't care so much about it. It's because they are a software company. In that sense, they are more like us than anyone else we have ever competed with." The seminal article and other commentary at the time asserted that Google's bypassing and marginalizing of Windows and other Microsoft software was not just something for the future but had in fact already started to become reality.

The final paragraph of the lengthy, thoroughly reported story had a comment from one of the most highly respected tech-industry analysts, Safa Rashtchy of the Piper Jaffray securities firm. "Google is a huge brand. From where I sit, it's their game to lose." A companion piece in the magazine, headlined "Living in a Google World," summed up the sentiment of technology industry experts analyzing the two companies and their respective trajectories this way: "Welcome to Microsoft's nightmare."

That afternoon in Seattle, Eric Schmidt ended his pugnacious presentation and opened the floor for questions from faculty and students at the University of Washington. He wasn't sure what to expect, given the university's physical proximity to Microsoft and the likelihood that some of the computer scientists at the gathering probably had worked or interned for the firm, or had their tuitions paid for by corporate grants. Of course, Google itself had hired more than 40 undergraduates from the school and at least 15 Ph.D.'s—a fraction of those on staff at Microsoft, but a meaningful number in a short period of years. Schmidt also scanned the room with interest, certain that Microsoft had at least one spy on hand to capture his every word, and to take note of the audience's receptiveness to his message.

The first question, about the size of project teams, was an easy one. Google did its best work on numerous initiatives in teams of three to five people, with their leader dubbed UTL, for Uber Team Leader. "We try to keep it small. You just don't get productivity out of large groups," Schmidt said. The students were lobbing him softballs: a follow-up question dealt with the pace of new product introductions. "Our rate of innovation is significantly faster than anyone in the industry and significantly faster than our competitors," he said bluntly. Knowing how much computer scientists loved their independence, Schmidt could hardly wait to answer the next query, which dealt with the management of technologists. "The secret here is not the way we manage but in our selection of the people," he said. "This model works when you have the right people. It would be a complete failure in an organization of people who wanted to be told what to do and had one big project. We try to have as little middle management as possible. They get in the way."

Given the brainpower in the room, Schmidt knew it was inevitable that he would end up having to duck some questions, like the one that came next. Asked why Google didn't provide more details about how its business really made money, so that analysts could build models and project its future profits and determine if its stock was priced correctly, he demurred. "We have decided not to give that information out," he said. "We don't want our competitors to learn it. We have invented models they don't have, which is one of the things that is driving things so quickly from a revenue perspective." Then he got another layup. Why was it, a student asked, that Google didn't seek to make money from its popular Google News Web site? Was it because it was using free content from other news providers? Just what was the reason that there were no ads? "Google News is a very successful product for us. Every month or two we have this conversation," Schmidt said. "Google News is easily monetizeable and would make lots of money. The team comes in and asks, 'Do you want more revenue or do you want Arabic news?' That seems like a no-brainer. We want Arabic news. There are only about 300 countries. Eventually we will run out of this and then we will work on monetization."

Now, predictably, came the harder questions. Asked what social impact Google's gathering and dissemination of the world's information would have, Schmidt took a breather. "We don't know," he said. "I personally believe the right model is to think of all the world's information in the equivalent of an iPod. What happens when you are carrying all that information with you and there is a real-time update? What does it do to teaching when every student can do the answer quicker than any professor can get it out of his or her mouth?" Then came a tougher query. Google, a student said, had no products to sell and few registered users, and if someone came along with a better technology for search, it could lose its following. Wasn't this a risky way to run a business? "Our business has relatively little lock-in," Schmidt conceded. "It means our competitiveness is search by search, person by person. We understand that and talk about that a lot. We have to be better on every search in every language, everywhere in the world."

Next someone played the Microsoft card, asking about the potential for getting caught in the cross fire of government regulators as Google grows larger and more powerful, and competitors had an incentive to spur officials in the U.S. and elsewhere to take action to rein in the company's runaway growth. It had happened to Gates. Couldn't it happen to Google? "It is a problem to worry about," Schmidt acknowledged. "We are a big enough force now that people pay a lot of attention to us. We are creating some negative waves for people who are disenfranchised or unhappy or don't want us for some other reason. I think it is going to get worse as we become more and more of the fabric of the online experience."

Schmidt didn't want to end the session on such a dour prospect; the students might leave with the feeling that Google could become the Justice Department's next antitrust target. Its very motto, Don't Be Evil, was a thinly veiled way of letting the technologists of the world know that Larry and Sergey were not just the Google Guys, but the Good Guys, who did the right thing for users and for employees, and had fun too. They weren't out to destroy competitors like the predators described in the massive litigation against Gates and Microsoft, Schmidt implied. He was fond of saying that Google had no real enemies.

As luck would have it, Schmidt was asked about the company's role as a gatekeeper of online information for Web sites that might want to charge for their content. Though he generally refrained from talking about products in the pipeline, Schmidt saw this as an opportunity to take the students into his confidence and make them feel like Google insiders. It was a classic, traditional recruiting technique he had used successfully before. "We have built and are in the process of building what we call proprietary content distribution," Schmidt said. "The vast majority of the money goes to the content provider. We keep a fee for providing the billing relationship." Then he expanded on the firm's role as the gateway to the Internet. "Google is likely to be, if not the, one of the largest places people come from, from the standpoint of being an information publisher. So you will want to work with us. We hope we would be the largest traffic generator to you as a publisher. We think that is good."

CHAPTER 24
Money Machine

As Google's stock price climbed, a major question loomed among investors: Was this a great stock to buy or a dangerous high-wire act? Wall Street commentators and experts cautioned investors to keep their distance. Google the search engine was everywhere, but Google the stock was difficult to understand, an enigma of sorts. Away from the herd on the Street, however, an independent-minded investment guru had other ideas. Baltimore's Bill Miller ran the vaunted Legg Mason Value Trust, a mutual fund that had outperformed the stock market for 13 consecutive years, putting him in a league of his own. A portly man nearly twice the age of Google's founders, who had once studied philosophy, Miller looked for young companies with sustainable competitive advantages that would lead to enormous growth and profits in the long run. Then he made big bets on those firms.

When Miller looked at Google's financial statements in the walk-up to the IPO, he saw a high-octane money machine powered by a search engine. For a young company, Google was a big-time moneymaker that had competitive advantages and enormous long-term potential. It already was earning hundreds of millions of dollars annually, its sales were skyrocketing, and it was in its infancy. The firm was getting bigger and bigger, faster and faster. Google had made so much money so quickly that it had no debt.

Even sweeter, its profits depended on Internet advertising, an expanding pie. As people spent more time online, businesses sought to reach them on the Internet.

"We were delighted when the so-called FUD factor"—fear, uncertainty, doubt—"dominated the process, resulting in the shares coming at the bargain price of $85, valuing the company at $23 billion roughly," Miller said. "There was the news flow about the deal: how the founders were arrogant and inexperienced, how the dual class share structure smacked of bad corporate governance, how the company would not provide enough information about its business, how it refused to give guidance on future prospects and so on." The noise drove the price of Google's IPO down, but it had nothing to do with the long-term growth and profit potential of the business. In fact, as Miller said, "We believe this created an opportunity."

Under Miller's direction, Legg Mason had bought more than four million Google shares in the IPO, spending hundreds of millions of dollars to amass a big stake. No stranger to the Internet, Miller's celebrated fund already had major holdings in Amazon.com and eBay, two best-of-breed firms that would soon celebrate their tenth anniversaries as dot-com stars. For the average investor who couldn't read a financial statement, one of the easiest ways to make money in stocks was to copy the moves of an investment genius. For anyone who wanted to know whether to purchase Google stock soon after the IPO, Miller's mammoth purchase, publicly disclosed in the fall of 2004, was a screaming "buy" signal.

While others scratched their heads, Bill Miller had confidence in Google's ad model. Advertising was undergoing an enormous shift, as billions of dollars moved from traditional media to the online world. And Google, more than any other company, had figured out how to profit in multiple ways in this new landscape, making it a technology company with a robust business. Broadcast television, magazines, and newspapers had for decades been fueled by advertising, and had been great moneymakers. Google was no different, Miller reasoned, just because this was the Internet.

Many wrongly associated Google stock with the high-flying firms of Silicon Valley's boom-and-bust era several years earlier. Back then, some Internet firms reported plenty of ad revenue too, but that was mostly from ads purchased by other Internet firms. In contrast, Google's ad dollars came primarily from thousands of small and medium-size businesses, many of which hadn't advertised on the Web before. These were a mix of online enterprises and traditional bricks-and-mortar firms. Amazon and eBay both were big advertisers on Google, buying targeted ads that would send thousands of computer users to their well-established Web sites. Google was also avoiding the kinds of ads that computer users hated; it was profiting instead from text ads triggered by searches. In a sense, what it had mastered was the reverse of mass-market advertising. Searching and surfing the Web with Google-powered ads was akin to driving down a highway and seeing only billboards that directly related to what you were thinking or discussing at that moment.

The scale of Google's business model, with the world's best hardware and software fueling it, was impressive. Even as it grew, the company was signing up new advertisers almost entirely through self-service over the Internet. This cut costs, increased revenue, and magnified opportunities for everyone involved—advertisers, Web publishers, consumers, and Google too.

Another competitive advantage Miller liked was Google's brand awareness. No other company in history had achieved this level of global recognition without spending heavily on advertising and marketing. Moreover, Google's network of affiliated Web sites created an advantage that would be hard to replicate: with thousands of Web sites, including AOL, *The New York Times,* and Univision, signed up to carry a Google search box, its name and logo were emblazoned in prime spots. As this network continued to grow, so would the brand.

In fact, with thousands of large and small Web site publishers running ads served up to them hassle-free by Google, the search firm had effectively become an online ad agency. On the affiliate sites, these ads were often but not always labeled "Ads by Gooooooogle." Google shared a generous portion of the revenue from these ads with its affiliates, creating among them a strong vested interest in the search engine's continued financial success, since it meant they would profit too. Yet, while Google sent monthly checks to these publishers, the company declined to share information about how it calculated what to pay them. For competitive reasons, it also refused to provide hard data about the performance of specific ads, leaving the publishers with little choice but to trust Google or opt out of the lucrative network.

The network was so lucrative, however, that many happily participated. And at the top of this list was Ask Jeeves. Reflecting the potency of the Google Economy, the value of Ask Jeeves soared, and by 2005 it became a $1.86 billion acquisition target. Nearly all of its revenue was tied to Google's ability to sell ads to businesses and place them on the Jeeves search engine.

Still, millions who loved the search engine continued to have a hard time understanding how Google made money, since they could use it for free. Many failed to grasp the difference between the free search results and the ads that appeared alongside them. Even among users who understood the distinction, many clicked on the ads so infrequently that they couldn't fathom how Google generated billions of dollars of revenue—especially since each click's value was often measured in pennies and not dollars.

Like so much about Google, it came down to sheer math. At the scale at which Google operated, serving results for hundreds of millions of searches each day, all it took was one out of every ten or fifteen searchers to click on an ad—at, say, an average price of 50 cents for the click—for the company to reach the sort of quarterly earnings it achieved in 2004.

If average users had a hard time grasping how Google earned its billions, Wall Street analysts remained perplexed by the company's unconventional methods. Here was a company that provided millions of computer users around the world with answers to their queries, yet in certain respects shrouded itself in secrecy.

It deliberately chose not to follow the crowd by giving analysts guidance about future products and quarterly profit. Sergey and Larry may have been forced to take the company public, but that did not mean Google was going to give out information that offered competitors clues about its future strategy. Instead, the trio at the top repeated with frequency that Google would be opportunistic as it expanded while also remaining faithful to its mission.

Analysts were left to grapple with questions on their own: What kind of quarterly earnings would Google report, especially since the company had said that it was managing strictly for the long-term? How would the market absorb all of the additional shares floated during the first six months of trading? And what about the threat posed by Microsoft? In addition, Google stock was not immune to something called "the law of large numbers." As the company grew, even if it remained extremely profitable, there eventually would come a time when its expansion would slow. But like its founders, Google remained young and hungry, just the kind of stock Bill Miller loved.

During the first six months of trading after the IPO, millions of additional Google shares were scheduled to become available for sale by insiders, putting downward pressure on Google's stock price and potentially flooding the market. On Wall Street, they referred to this as an "overhang" of shares, and it raised the prospect of choppy trading and wide swings in price. However, Google stock proved resilient and climbed steadily, hitting $135 in October, a mere two months after the company went public at $85. Nonetheless, several Wall Street analysts urged investors to sell, calling this abrupt rise in share price a speculative frenzy.

"There may be no near-term support for the stock on any correction," said Mark Mahaney, an analyst with American Technology Research. The Google bubble would burst due to "overly aggressive expectations" about the company's quarterly profits, he warned. But after Google reported healthy quarterly sales and profits on October 22, the stock rose again.

Defying the law of supply and demand, this price rise continued. To start the New Year off right, on January 3, 2005, Google stock closed above $200 for the first time, another milestone. On February 1, the day after the company reported extraordinarily strong quarterly sales of more than $1 billion and profit of more than $200 million, the stock hit $216.

Google's stock market value now exceeded $50 billion, making it worth more than many of the biggest and most prestigious corporations in America. The sales and profits that it was producing were off the charts, propelling the stock to new heights based on expectations of continuing strong performance. In the short run, however, one major hurdle remained: an enormous overhang of some 177 million shares of additional Google stock slated to be released for sale by employees and other corporate insiders on February 14, Valentine's Day. Even for Google, this was a lot; it was more shares than the company had outstanding, and it would bring the total number of Google shares available for trading to about 300 million. In anticipation of the overhang, Google's stock price began declining in the weeks leading up to February 14, falling below $200 and leaving many shareholders wondering if they had been lucky so far and ought to sell, or if this was a temporary pullback.

There were other pressures too. Google executives had disclosed plans to sell millions of shares of stock that they owned. All of this had been reported publicly and was the subject of endless debate and discussion in the financial press. The selling made investors nervous: they didn't want to be buying if Larry and Sergey were selling. That was true even if the founders still owned nearly all of their original Google stock and were merely selling to diversify their holdings. At these lofty price levels, some Wall Street skeptics pinned a derogatory label on Google. They called it a one-trick pony. By this, they meant that the company earned all of its money from a single source: Internet advertising tied to searches. But Wall Street analysts had to keep revising their Google share price targets upward as the stock rose, even as they complained that it was impossible for them to get a clear idea of the real value of Google stock because the company steadfastly refused to provide them any guidance.

On February 9, 2005, Google opened its doors for the first time to Wall Street financial analysts, who traveled to Silicon Valley for the chance to meet Larry and Sergey and the executive team. This was only days before the additional 177 million Google shares were to become available for trading, and the goal of the session was to provide more information at what CEO Eric Schmidt recognized was a critical juncture. He understood the importance of Google opening up a bit, and of making it through Valentine's Day without a stock market event that might shake confidence in the company.

The day opened with reassuring words from Schmidt. "The advertising network is a beautiful thing to behold," he said. "We have a wonderfully diverse set of advertisers. We're not reliant on any particular category or advertiser to some overwhelming regard.

"Part of it is because of this concept called 'The Long Tail,'" he went on. This was the idea that in the Internet age, geography mattered less as low-cost distribution enabled niche products, catering to very specific tastes, to attract large audiences. It turned out that the most popular books, music, and movies made up a surprisingly modest proportion of sales for Amazon, Netflix, and other retailers; the rest came from a "long tail" of obscure favorites that the Internet had made easier to find. In Google's case, the concept applied to the unexpectedly wide array of businesses that bought advertising on the search engine.

"The surprising thing about The Long Tail is how long the long part of the tail really is, and how many small businesses there are that have not [had access to] the mass market," Schmidt said. "We're doing pretty well in the middle of the tail. We still do not have all the products and services that in our judgment are really needed to serve the largest advertisers, or the very tiniest of advertisers. We want to make sure we serve this whole space very, very well, and that is all stuff under development."

Schmidt's words made it clear that Google's advertising and business model had plenty of room for growth, and also that the firm planned to go after deep-pocketed Fortune 500 advertisers in 2005. "We are not quite as unconventional as we actually say all the time," he said. "The things that we do are unique in the way that products are created, but much of the rest of the business is run in all the normal ways and very much at the state-of-the art but in a traditional way. We actually do care about objectives. Every quarter we go through, 'How are we doing?'"

So Google did care about its financial results after all. It was simply that with so many brilliant engineers and mathematicians, it went about the process of innovation in ways that were more like a university than a business. As for management and financial resources, Schmidt explained they were allocated 70/20/10. By that, he meant that 70 percent was put into core search and advertising, 20 percent was put into adjacent products, and 10 percent was invested in totally new ideas, the vision of things to come. "You have to work on your core business because that is the thing that brings in the cash flow, the customers, the business," he said. As for the 10 percent, "We have a very sophisticated set of product managers and technology leaders who understand how to take some of these brilliant ideas and turn them into businesses."

Having heard precisely what they needed to hear from Schmidt, the analysts now waited for the founders to offer their views. Notwithstanding the reassurance from Schmidt, the guys remained the company's controlling shareholders. More than anyone, they could influence the future direction and management of Google.

Brin said he focused on motivating and retaining the best and brightest from around the world. Now that Google was public, new financial incentives were needed. To foster innovation, he said, Google was creating "Founders' Awards," multimillion-dollar stock awards to be presented regularly to small teams of people that developed the best new ideas. Big money like that was unheard of at most companies, and its goal, he said, was to retain brilliant innovators who might otherwise leave, taking their ideas with them.

Page said he spent his time on innovation and on improving existing products. An enormous quantity of computational resources was devoted to delivering terrific search results and relevant ads to hundreds of millions of people at the lowest possible cost. "We're trying to be ruthlessly efficient about how we run our business, and we are in this to make a lot of money," he said, "but we're not necessarily going to make money from all the things we have."

The analysts departed satisfied. Schmidt was a pro, Brin and Page were grown-ups, and Google stock would continue to rise. It didn't seem to bother the analysts that Brin, Page, and Schmidt had taken some of their chips off the table by selling hundreds of millions of dollars of Google stock. They were entitled to diversify their investments, and in any event, the founders still owned billions of dollars worth of Google stock.

By the time the company held its first annual meeting for shareholders, on May 12 at the Googleplex, its stock price had passed $225. A few weeks earlier, it had reported fantastic results for the first three months of 2005. Profits had soared 600 percent to $369.2 million and sales reached $1.3 billion. The company served lunch, including Snicker-Google cookies, to the few hundred shareholders who showed up at the session. Others—including members of the media, who were barred from attending—tuned in to an Internet simulcast.

The first person in line to enter the meeting was Jeff DeCagna, a Google stockholder from Washington DC who worked as a consultant. He was impressed by the way Google controlled every aspect of the session and made sure that stockholders could not wander the campus. Any one of them could have been a spy from a competitor. "I will continue to buy and hold," DeCagna said of Google stock. "It's cheap even at $200 in my opinion. Great companies believe and invest in innovation because it's intrinsic to their success. If they continue to follow the management style, I think it could go to one or two thousand a share."

By June, Google stock was the talk of the town. It was flirting with $300 a share, making the company worth more than $80 billion, and overshadowing the honor bestowed on Larry and Sergey when they were named to the prestigious American Academy of Arts & Sciences. On the cable television network CNBC, which tracked the stock market daily, the company's stock price was displayed, front and center, alongside the Dow Jones Industrial Average. The world was watching to see when Google, which had gone public at $85 less than a year earlier, would hit $300. "No company is as popular as Google," the *Financial Times* wrote in an editorial. "It would take only a tiny mishap or flattening of advertising growth for the shares to plummet. Does that matter? Probably not, as long as Google remains, as it shows every sign of doing, a proudly independent company. But anybody who takes its lofty valuation too seriously is in danger of making a big mistake."

In the week leading up to the Fourth of July, Google set off some fireworks of its own, breaking through the $300 barrier. Mark Mahaney—the Wall Street analyst who had advised investors to sell Google stock after it hit $135 in October 2004—was now bullish on Google, and employed at a new firm. From his post at Citigroup Smith Barney, Mahaney set a new price target of $360. Other analysts also pegged Google at prices well above $300. Like the search engine that powered it, Google stock had taken on a life of its own.

CHAPTER 25
The China Syndrome

As 2005 wore on, Bill Gates and Microsoft prepared to launch a major new offensive to counter Google's runaway growth, momentum, and success. The company's leaders chose their target with enormous care. There were many possible areas to consider. For example, there was Google's involvement with the funding of a new way to surf the Internet through a faster, more secure technology tool called Firefox. A direct competitor of Microsoft's Internet Explorer, Firefox was proving so popular that tens of millions of people downloaded it onto their computers and began using it to navigate the Web. Aside from Firefox, there was the issue of recruiting. Google was regularly hiring talented Microsoft technologists as if Gates were running some kind of employment agency for Larry and Sergey. This had to be stopped forcefully, and in a manner that would garner global attention.

There was perhaps no more important recruiting battlefield than China, which had the world's fastest growing economy and more than 100 million Internet users. China offered vast potential for new talent and profit through technology, and Microsoft was already well-entrenched there, with nearly 1,000 employees. Now Google wanted to establish its own product development and research center in China. For Google, China mattered. The country would eventually become the biggest Internet market in the world, and in terms of users, it was already the second biggest after the U.S. If Microsoft caused severe problems for Google related to China and recruitment, and played the spoiler, the ripple effects could be tremendous.

To succeed, Microsoft would have to change the terms of engagement. Every time the software giant had tried to counter Google's global growth in recent years with new products, it had stumbled. Google seemed to be what everybody was talking about in the world of technology, a sign that Microsoft was falling further behind in the battle for mind-share. It wasn't faring any better on Wall Street, where Microsoft couldn't get its stock price moving or garner the buzz that accompanied Google's every move. It was as if Microsoft were on a corporate treadmill, running in place. To be sure, the company still enjoyed enormous advantages in size—its $275 billion stock market value was more than triple that of Google's. It also had clout and recurring revenue from the sale of millions of personal computers loaded with its Windows operating system and suite of office programs, including Microsoft Word. It had a war chest of billions of dollars, and a diverse array of products, including popular new electronic games.

But in a world where perception mattered, Microsoft was viewed by many technologists as the Soviet Union of software, a lumbering giant unable to keep up in the fast-paced digital millennium. As a mature business built around the personal computer, Microsoft lacked the sex appeal that made Google the darling of the Internet. Gates desperately wanted Microsoft to be more than a fat cash cow. He wanted it to get its edge back, if not initially through innovation, then through brute force. A high-profile attack on Google that involved recruiting had the potential to garner publicity and maximum impact. And if it involved China too, it would be a story with all the elements.

Microsoft's main objective was to stem the flow of talent to Google, and in the process cut off the source of its sizzle. Gates convened a committee inside Microsoft that focused exclusively on devising ways to compete with Google. The group gave top management a confidential briefing called "The Google Challenge." Lately, for engineers, it had been a one-way street from Microsoft to Google. As thousands of résumés a day flowed into the search engine company, Microsoft struggled to keep its best people, even after offering them more money and perks. The brain drain of software mavens was unlike anything it had ever experienced. To make matters worse, Google was hiring Ph.D.'s from the top universities across the country and outrecruiting Microsoft in its own backyard at the University of Washington. Google's newly established outpost in Kirkland, just down the road from Redmond, was teeming with talent that had formerly worked for Gates; it was now easy for engineers to jump ship without having to move to a new city or even change their commute. Unable to talk his employees out of leaving, Microsoft CEO Steve Ballmer vowed, "I'm going to fucking kill Google." By hitting Google hard over its recruiting practices, and by improving its own hiring efforts on campuses and elsewhere, Gates and Ballmer hoped to make Larry and Sergey regret their aggressive poaching.

Something definitely had to be done. In the spring and summer of 2005, Google had stepped up the pace of its hiring, adding more than 700 employees in just three months. With Larry leading the recruiting effort, Google's total headcount soared to 4,183, nearly double the total a year earlier. The company also opened new offices around the world, including in Sweden, Mexico, and Brazil, and employed people in two dozen countries to lead these efforts. From Europe to South America to Asia, Google was on the move.

And then there was China. "China is obviously a very exciting market in general and also for Google," Page said. "We have actually a very significant market share in China. There's tremendous opportunity for us there with our existing market share to make money through advertising. We're just ramping up our business operations." After Google registered for an official business office in China, and CEO Eric Schmidt took a trip there to look matters over, Microsoft prepared to slam its nemesis hard.

Dr. Kai-Fu Lee began working for Microsoft in China in 1998, the same year that Google was founded. The highly regarded Ph.D.

from Carnegie-Mellon University, who maintained strong ties to the technology communities in China and the U.S., founded Microsoft Research Asia, a lab in Beijing that quickly developed a strong track record. In 2000, Lee transferred to Microsoft's U.S. headquarters, where his assignments ranged from overseeing the company's strategy in search to making it easier for computer users to utilize Microsoft products. He participated in meetings of the China Redmond Advisory Board, which focused on the status of Microsoft's Chinese operations and strategies, and assisted with finding vendors in China when necessary. He also met with Bill Gates to discuss issues posed by Google and search technology. Microsoft compensated Dr. Lee handsomely for his work, including paying him more than $1 million in 2004 alone.

Then, in spring 2005, Lee heard that Google wanted to open a major research operation in China, and he began talking to top company officials about the job of leading it. This would be an opportunity to build something from the ground up, and that kind of challenge appealed to him. So did the prospect of joining Google. As the talks progressed, Lee, who was still employed by Microsoft and had signed a one-year noncompete agreement, told officials there that he wanted to leave. Google, he explained, was offering to make him head of its new China operations.

If Lee left, he would become the highest-ranking employee snatched by Google from Microsoft, and a prime target for an all-out legal assault. In a sworn statement filed in the litigation that followed, Lee alleged that Microsoft senior vice president Rick Rashid warned him, "You should not go. Things will be very unpleasant for you if you go."

"If you leave, we would have to do something, and when we do something, please don't take it personally," Microsoft CEO Steve Ballmer told him, according to Lee's statement. "We like you. Your contributions to Microsoft have been immense. It's not you we are after, it is Google."

Finally, Bill Gates met with Lee, tried to persuade him to stay, and was completely direct about what would happen if he defected. "Kai-Fu," Gates told him, "Steve is definitely going to sue you and Google over this. He has been looking for something like this. . . . We need to do this to stop Google."

Despite the warnings, in July 2005 Lee followed the others who had left Microsoft for Google. The search engine giant hailed his recruitment as a coup that would enable the company to vault ahead in China. Lee himself said Google's expansion plans could lead to technology breakthroughs that would fuel the continued growth of both the Chinese and U.S. economies. His title, president of Google China, reflected the prestige he brought to the post.

Every declared war begins with an opening salvo of some kind. This one would be no different. It was Microsoft's turn to sully a Google announcement by firing a shot heard 'round the technology world. In a high-profile lawsuit against Google and Dr. Lee, Microsoft alleged that Google knowingly and improperly sought to induce him to violate the terms of his employment contract with Microsoft. The contract included noncompete and nondisclosure clauses that restricted Dr. Lee's activities, prohibiting him from using information learned at Microsoft to benefit Google's foray into China or its development of search technology.

"Dr. Kai-Fu Lee—with Google's encouragement—is blatantly violating his noncompetition promises to Microsoft. He is doing so by defecting Microsoft for Google, a direct competitor in markets as to which Dr. Lee holds Microsoft's most sensitive technical and strategic information," the company alleged in court papers. "Microsoft is engaged in intense competition with Google. By virtue of his leadership roles, Dr. Lee learned Microsoft's most sensitive technical and strategic business secrets about search technologies. He was also deeply involved in Microsoft's efforts to expand its business in China and Microsoft's confidential strategic plans regarding that crucial new market." While Microsoft was seeking an injunction blocking him from undertaking the assignment, Lee went immediately to China, before a U.S. judge ruled. He met with media there to get things rolling, as Google fired back at home.

"In a shocking display of hubris, Microsoft has rushed into

court claiming the entire field of search as its own," Google alleged in court papers. "It seeks to impugn Lee's good reputation with no evidence. The lawsuit is a charade. Indeed, Microsoft executives admitted to Lee that their real intent is to scare other Microsoft employees into remaining at the company. Microsoft's market domination eliminates any legitimate interest in protecting itself from competition through this lone employee."

Microsoft's lawsuit painted Google as a reckless corporate crusader with no respect for the rule of law, including the honoring of standard employment contracts. "Google is fully aware of Lee's promises to Microsoft, but has chosen to ignore them, and has encouraged Lee to violate them," the company alleged. "Were he to accept the position with Google, he would necessarily be helping Google compete against Microsoft's business strategies in China—strategies that he helped develop on behalf of Microsoft."

Despite Google's assertion that Lee was not a search engine expert, a judge in the State of Washington issued a preliminary ruling in Microsoft's favor, prohibiting Lee from engaging in work that involved search or Google's plans for China. The judge subsequently allowed Lee to begin hiring in China, but put limits on what else he could do. Google said it would abide by the rulings, pending a trial.

The temporary triumph over Google sent a message to technologists around the world. It raised the specter of litigation for any senior Microsoft employee who left for Google. Heavy press coverage in the U.S., China, and Europe depicted the struggle as an illustration of how seriously the software giant viewed the threat posed by its smaller search engine rival. "Clearly, Microsoft's motives reach far beyond Lee. By going after Google and Lee, it may help deter other talented techies with similar ideas," said one story in *BusinessWeek*. In China, *The People's Daily* offered another perspective: "Lee said his move is in accordance with moral norms and law. He holds one should persist in seeking his choice and his enthusiasm. Of course one should not violate law. Lee believes that his approach agrees with such principles."

Lee himself outlined the reasons for his wanting to join Google on a Chinese Web site read by thousands of engineering students. Formatting his statement as a math equation, he wrote: "youth + freedom + transparency + new model + the general public's benefit + belief in trust = The Miracle of Google."

In August 2005, amid the battle with Microsoft over China, Google celebrated its first birthday as a public company by surprising Wall Street anew: it announced plans for a $4 billion mega stock offering. The billions in fresh cash would give it a deeper war chest to fight the rising competition from Microsoft and Yahoo in the U.S. and abroad. In a whimsical touch, the search engine said it would sell 14,159,265 additional shares to raise the money—a number representing the first eight digits after the decimal point in pi, the mathematical ratio of a circle's circumference to its diameter. (In its IPO a year earlier, Google had based the total amount of money to be raised—$2,718,261,828—on another mathematical figure known as "e.") To Google-watchers, this was a sign that the company, despite its newfound riches, had retained its quirky corporate culture. The company also announced it would hold a cook-off at the Googleplex to select two new executive chefs to replace Charlie Ayers. "We welcome all culinary engineers to try out," Brin said.

Larry and Sergey, mathematicians turned moguls, were still turning the ordinary into the extraordinary. In the months following Google's IPO, the 31-year-old founders became America's newest and wealthiest young billionaires, and hundreds of other Googlers became millionaires. The guys debuted on the Forbes 400 list of the wealthiest Americans at number 43, with a net worth of $4 billion each. That was based on a Google stock price of about $110. When the stock shot up in price to more than $300 in the summer of 2005, each of the founders had a net worth in excess of $10 billion. Still, this wasn't good enough for Sergey's mother, Eugenia Brin, who wanted him to return to Stanford, write his thesis, and finish his Ph.D.

Sales of Google stock by its officers, directors, major investors, and employees approximated $3 billion within the year. But the selling by Larry, Sergey, and the others did little to dampen investor enthusiasm for the company. Often, selling by top corporate officials is perceived negatively on Wall Street, but Google defied that trend too. In addition to stock they sold in the IPO, Larry and Sergey unloaded about 400,000 shares per month, providing each of them with more than $750 million in cash, and CEO Eric Schmidt sold about 113,000 shares every month, giving him more than $225 million. (When details about Schmidt's financial holdings and other personal information appeared in the online technology publication CNET, Google, protective of the privacy and safety of its top executives, said it would not talk to any CNET reporters for one year. Many commentators accused Google of hypocrisy, given the company's emphasis on making abundant information about people, places, and things easily available online, and also because CNET had used Google itself to unearth the information.) Early investors John Doerr and Ram Shriram cashed in too, selling $45 million and $313 million of Google stock respectively. And Stanford president John Hennessey unloaded stock he received for serving on the board of directors, raising $2.5 million. Given Google's lofty valuation, they all recognized that any number of things could go wrong and send Google's stock price plunging.

In deciding how to handle their stock sales, Larry and Sergey wisely heeded the advice of financiers and lawyers who had seen booms and busts for years. The founders planned to hold on to most of their shares, but they did not want to end up like those Silicon Valley entrepreneurs who fell in love with their company's stock, never sold any, and eventually had nothing when the business fell apart. So whether Google's stock price was rising or falling in any given month, it made sense for each of them to sell the identical number of shares on the same date on a regular schedule. This avoided two problems. First, since the selling was on autopilot, there would never be any question about whether they bought or sold stock based on confidential, inside information. Second, it would cause them to convert a portion of their stock holdings to cash, ensuring that whatever happened, they and their families would have more money than they would ever need, as well as diversification of investments. And, because of the existence of two classes of stock with unequal voting rights, they could safely unload stock without giving up control of Google's destiny.

Many Googlers, including the founders, used cash from stock sales to buy new houses. Omid Kordestani, head of sales, sold hundreds of millions of dollars of stock and broke into the limelight in a front-page *Wall Street Journal* story that reported he had paid $17.8 million for a 16,000-square-foot house in Atherton, a suburb near Google's Mountain View headquarters. A number of other Google employees joined Kordestani by purchasing houses in the Atherton area, where residential real estate values were at such breathtaking heights that its postal zip code, 94027, famously became among the most expensive in the nation. Googlers also purchased houses in Menlo Park, where Larry and Sergey had launched the company in a garage, and in Palo Alto, near Stanford's campus. To win clients, real estate agents in the area advertised on Google.

Flush with $625 million in cash generated in the second quarter of 2005, Google went shopping too. The company invested several million dollars in Current Communications, a private company based outside Washington DC that was in the early stages of providing high-speed Internet access over standard electrical power lines. Most existing high-speed Internet access in homes was provided over cable television wires or telephone lines. The investment reflected Google's strong interest in the cost and sources of power, a decisive variable in choosing locations for the data centers where the company houses its hundreds of thousands of computers. The investment was also part of bringing the Internet and Google to more people around the world.

As the Lee drama continued to unfold in the late summer of 2005, the Internet market in China heated up, further raising the stakes. Yahoo announced an investment of $1 billion in a leading Chinese

Internet firm, Alibaba, even as Microsoft's actions in the Kai-Fu Lee matter temporarily disrupted Google's own expansion there. But Google was nimble, and the company had multiple access points to the Chinese Internet market.

One of these was Baidu.com. In 2004, Google had purchased a 2.6 percent stake in the leading homegrown Chinese search engine, whose logo of primary colors on a pure white background was fashioned after Google's. The word *baidu* means "100 times" and is linked to an ancient poem about a man searching for his lover. In another indication of the Google Economy's astonishing scope, Baidu.com went public on Wall Street in August 2005, billing itself as "China's Google." While it had only $8.4 million in quarterly sales and $1.5 million in profit, the rub-off from Google and the prospects for China's rapidly growing Internet search market were all that it needed. On its first day of trading, Baidu stock soared from $27 to $122. This was the biggest first-day gain in an IPO since the dot-com bubble had burst five years earlier.

"This is a 'son-of-Google' investor mentality," said David Menlow, president of IPO Financial Network. "Everyone remembers they could have had Google at $85 and don't want to let it happen again."

The overheated Baidu IPO added billions of dollars of market value to the global Google Economy, which was growing elsewhere too. That same week, a British company paid $43 million in cash to buy Search Engine Watch, the trade show and online newsletter operation run by former newspaper reporter Danny Sullivan, who continued to carefully track Google's every move for his global audience of readers.

Google's own stock market value now neared $80 billion, making the company more valuable than Amazon and eBay combined. While Google thought of itself as an engineering and technology-driven firm, it made its money the same way most media companies did—primarily through advertising. The irony was that a company whose financial strength was built on advertising did almost no advertising itself. It didn't need to. During its first year after going public, Google was worth slightly more than the world's biggest media company, venerable Time Warner, despite that corporation's far-flung holdings from Hollywood studios to cable channels to magazines to America Online. It was also worth more than Disney, Ford, and General Motors, established U.S. companies with global reach and reputations. To appreciate how lofty Google's valuation had become, consider this: its market value was more than 25 times that of Dow Jones, publisher of *The Wall Street Journal*; about 20 times the value of The New York Times Company; and roughly 10 times the worth of The Washington Post Company.

To earn even more, Google worked on new payment methods to make online purchases easier for computer users and experimented with buying and reselling print ads in magazines. It also developed improved ways for advertisers to precisely measure and track sales generated from ad spending on Google, making the results of that advertising more quantifiable. It provided new ways for major corporate advertisers, including retailers Wal-Mart and Costco, to have greater control over the placement and timing of their online ads than in the past. And it reorganized its sales force to better serve the biggest U.S. corporations. Forging into new areas with big profit potential, Google technologists worked closely and quietly with Hollywood studios to develop methods to protect digital video rights so that movies could easily be accessed, downloaded, and paid for online, putting the company in direct competition with similar efforts by Microsoft. Still, maintaining Google's lead in search mattered greatly to Larry and Sergey. That was one reason why most of the searches on the Internet—nearly 60 percent in the U.S., according to survey data—took place through Google.

"We have figured out ways to stay focused on end users and innovation," said Eric Schmidt. "We're sticking to the focus that we talked about in the original founders' letter around getting all of the world's information online. And we have been able, even with growth and so forth, to be able to attract the very best and brightest around the world. We're very excited about the talent we have assembled, the scale that we operate in, the computing power of the technology that we have built here, and all of the innovation. The most direct way to access the world's information will be through Google."

CHAPTER 26

Googling Your Genes

Sergey Brin and Larry Page have ambitious long-term plans for Google's expansion into the fields of biology and genetics through the fusion of science, medicine, and technology. Their goal—through Google, its charitable foundation, and an evolving entity called Google.org—is to empower millions of individuals and scientists with information that will lead to healthier and smarter living through the prevention and cure of a wide range of diseases. Some of this work, done in partnership with others, is already under way, making use of Google's array of small teams of gifted employees and its unwavering emphasis on innovation, unmatched search capacity, and vast computational resources.

"Too few people in computer science are aware of some of the informational challenges in biology and their implications for the world," Brin says. "We can store an incredible amount of data very cheaply."

He and Larry want to make it easier for users to find the right information faster, and the company is pouring the bulk of its resources into enhancing the breadth and quality of search. This involves wholly different methods of searching that may eventually make today's Google seem primitive. As these evolve, the search mechanisms of the future will produce better answers to queries, just as Google is superior to the early search engines that preceded it.

"The ultimate search engine," says Page, "would understand exactly what you mean and give back exactly what you want."

The critical path inside the Googleplex includes experimentation with artificial intelligence techniques and new methods of language translation. Brin and Page are hopeful that these efforts will eventually make it possible for people to have access to better information and knowledge without the limitations and barriers imposed by differences in language, location, Internet access, and the availability of electrical power.

To assist in this effort, Larry and Sergey have recruited a diverse group of people to work at the Googleplex, including a collection of former CEOs, hundreds of Ph.D.'s, U.S. and world puzzle champions, former Olympians, an award-winning independent filmmaker, and a coterie of university professors.

Brin and Page foresee Google users having universal access to vast repositories of fresh information, some of it public and some private, which is not currently available on the Internet. This encompasses motion pictures, television, and radio programs; still images and text; phone calls and other voice communications; educational materials; and data from space. The pair is also involved in the hunt for clean, renewable energy sources to power Google and broaden economic growth. "These guys have a big, compelling vision for what the company is going to do," said Stanford president John Hennessy. "They think very hard about the long term."

One of the most exciting Google projects involves biological and genetic research that could foster important medical and scientific breakthroughs. Through this effort, Google may help accelerate the era of personalized medicine, in which understanding an individual's precise genetic makeup can contribute to the ability of physicians and counselors to tailor health care treatment, rather than dispensing medications or recommending treatments based on statistics or averages. New insights, new medicines, and the use or avoidance of certain foods and pharmaceuticals for people with specific genetic traits are among the possible outcomes.

"Just think of the application of Google to genomics," said Hennessy. "There are large databases, lots of information, and the need for search." With the addition of specialized data, he said, Google's index could aid in new discoveries in genetics. "You want to be able to use a search system that is content-dependent, with the genome and structure of DNA already built in. It is one of many potential areas where you can see this so-called 'intelligent search' making a big difference. We are going to see more and more of it."

Dr. Alan E. Guttmacher, deputy director of the National Human Genome Research Institute, said Google's involvement in genetics is particularly meaningful because of its capacity to search and find specific genes and genetic abnormalities that cause diseases. He also said that its massive computing power can be used to analyze vast quantities of data with billions of parts—quantities that scientists in laboratories do not have the capacity to process. The old model of a scientist working in a lab, he said, is being replaced by the new paradigm of a researcher working at a computer, connected to databases through the Internet, and doing simulations in cyberspace. "Until recently, the challenge has been gathering data," Guttmacher said. "Now, the bigger challenge is organizing and assessing it. Google-like approaches are the key to doing that. It completely accelerates and changes the way science is done. We are beginning to have incredible tools to understand the biology of human diseases in ways we never have before, and to come up with novel ways to prevent and treat them."

Over dinner and plenty of wine in February 2005, Sergey Brin discussed the prospects for genetics and Google with the maverick biologist Dr. Craig Venter. Venter, who had decoded the human genome, was in the midst of gathering oceanic samples from around the world and sending them back to the U.S. for analysis of nature's DNA. Despite millions of dollars in funding and thousands of hours of computing time from the federal Department of Energy, Venter needed more help to unlock the molecular mysteries of life. It seemed to him that Google's mathematicians, scientists, technologists, and computing power had the potential to vault his research forward. He pressed Brin hard to get Google involved.

Also at the table was Ryan Phelan, chief executive officer of DNA Direct, one of the foremost Internet companies providing individuals with genetic testing and counseling. DNA Direct gets nearly all of its patients through ads it buys on Google. The ads appear to the right of the free search results when users type in "blood clotting," "breast cancer," "cystic fibrosis" or certain other diseases. Brin, Venter, and Phelan were among those who had been invited to a dinner of the wealthy and wise at Cibo, a trendy Italian bistro in Monterey, California. Brin had brought along his friend Anne Wojcicki, a health care investor whose sister is a senior executive at Google. Seated nearby was early Google investor Jeff Bezos, the CEO of Amazon.

"What [Venter] was talking about with Sergey was, 'How can you use Google to really help access everything at the genetic level?'" Phelan recalled. "What Craig was after was pure raw science. What I was hearing was, 'What if Google was the repository for the distribution of this information?' Sergey is so intellectually engaging. He was trying to pull out from Craig how you could use Google and how it could make a difference."

Google is not averse to contributing to the scientific efforts of others. It teamed up with Stanford several years ago to provide computing power for a scientific project that focused on unfolding proteins. The process of protein folding is one of the keys to understanding biology, yet very little is known about how it works. It is believed by some that when proteins fold incorrectly, it can lead to serious diseases, ranging from Alzheimer's to Parkinson's to many types of cancer. The Stanford project utilized idle computer time from the PCs of individual volunteers and organizations like Google that agreed to apply excess computational power to the gargantuan effort to simulate the protein-folding process in 3-D. Google also made it easy for individuals who downloaded its search toolbar to sign up for the Stanford program, so that while they were away or asleep their computers could be utilized in the cause of science. The extra computing power accelerated the simulation and analysis of protein folding. "Modeling even the simplest of proteins can be computationally very, very challenging," Brin said.

Not long after the dinner in California, Brin and Page teamed up with Venter. The biologist gained access to Google's immense computing power and personnel. He said this would accelerate analysis of molecular data and significantly increase the likelihood of advances in both applied health care and basic scientific research.

"We need to use the largest computers in the world," Venter said. "Larry and Sergey have been excited about our work and about giving us access to their computers and their algorithm guys and scientists to improve the process of analyzing data. It shows the broadness of their thinking. Genetic information is going to be the leading edge of information that is going to change the world. Working with Google, we are trying to generate a gene catalogue to characterize all the genes on the planet and understand their evolutionary development. Geneticists have wanted to do this for generations."

Over time, Venter said, Google will build up a genetic database, analyze it, and find meaningful correlations for individuals and populations. It is utilizing the 30,000 genes discovered by Venter and scientists from the National Institutes of Health when they were racing to beat one another to map the human genome. On June 26, 2000, federal researchers and those from the private sector came together at the White House to announce that their race to map the human genome had ended in a tie. Shortly thereafter, Venter and scientists from NIH made the genetic information they had gathered publicly available on the Internet, a stark contrast to the days when scientists hoarded data. Google went on to post a double helix doodle on its Web site to mark the fiftieth anniversary of the discovery of the structure of DNA, the material inside cells that carries genetic information.

Google's data-mining techniques appear well-suited to the

formidable challenges posed by analyzing the genetic sequence. It has begun work on this project, but has not been required to disclose any information about it publicly since the work has no impact on its current revenue and profits.

Brin, who has long had a serious interest in molecular biology, is deeply engaged by the role that Google can play in enhancing "the ability for cellular biologists and other kinds of medical researchers to be able to start to use data clusters like we have at Google, and certainly like the ones we're going to have in a decade or two decades' time, and be able to do completely new things that we weren't able to dream of before." The implications could be significant for individuals. While genetics does not necessarily provide yes-or-no answers to various medical and other questions, it does offer probabilities and statistics that can guide decision-making.

"People will be able to log on to a Google site using search capacities and have the ability to understand things about themselves as they change in real time," Venter said. "What does it mean to have this variation in genes? What else is known? And instead of having a few elitist scientists doing this and dictating to the world what it means, with Google it would be creating several million scientists.

"Google has empowered individuals to do searches and get information and have things in seconds at their fingertips," he went on. "Where is that more important than understanding our own biology and its connection to disease and behavior? With Google, you will be able to get an understanding of your own genes. Google has the capacity to do all of this, and it is one of the discussions I have had with Larry and Sergey. They are the right people to undertake this." Stewart Brand, a technologist and futurist who was also at the dinner where Brin and Venter discussed Google and genetics, described the alliance of these mavericks as "a match made in heaven."

Venter may have discussed matters with Brin first, but he bonded with Larry Page too. In April of 2005, Page invited Venter to join him as a member of the board of directors of a foundation encouraging a private space race. The X Prize Foundation is fashioned after the Orteig Prize won by Charles Lindbergh in 1927 for his New York-to-Paris flight. Its stated mission is to promote competitions to foster breakthroughs in space travel and related technologies. The foundation shares something in common with Google: innovation through small, highly motivated groups of bright people who are given access to immense resources.

A few weeks after Google went public in the summer of 2004, Mojave Aerospace Ventures Team, led by Burt Rutan and funded by Microsoft cofounder Paul Allen, flew the world's first private spacecraft to the edge of space and won the $10 million Ansari X Prize, cementing the foundation's model for innovation through competition. Upon joining the board in January 2005, Page said he was "excited to be working with the foundation to foster additional breakthroughs." Venter said he was honored and pleased when Page asked him to join the distinguished group.

"They're trying to foster competition to get people into space," Page explained. "I have a good friend who really wants to go to Mars, and so he decided he should build a rocket company. He has been pretty successful about it. I just sent him an email, and I asked him for some stats. 'So what is the theoretical cost of getting a pound of something into space?' It's basically the fuel that powers the rocket into orbit. The Space Shuttle costs about $10,000 to $20,000 per pound that it carries up. What do you think the theoretical lower limit is? It's actually about $10 to $20 per pound to move something into orbit. For you or your body, that's probably the cost of an expensive airplane ticket, right? Do you think someday we might figure out how to get close to that? I think we could. That would change things a lot and might get us to Mars."

Back on earth, Page also foresees greater involvement for himself and Google in causes that work to relieve hunger and poverty through entrepreneurship, self-reliance, and philanthropy. He has taken a particular interest in programs that provide small bank loans to people in developing nations. "Mohammed Yunus in Bangladesh has given out over $2 billion now, $160 at a time, to poor people and been very successful," he said. "The money gets returned and it is a functioning business." He added, "I believe eliminating poverty is something we could do. Bono is actually more eloquent than I am on this, so I'll read you a quote from him: 'Africa is not a cause. It is an emergency.'"

Among the other innovations that Sergey Brin and Larry Page would like to see Google and other firms achieve in the future is the production of affordable, clean-burning fuel that does not harm the environment. The source for this power is likely to be the sun. This area of research is important to Page, who for years has focused on the enormous quantities of electricity needed to power Google's network of hundreds of thousands of computers.

While it is possible that some of Venter's biological research may lead to discovery of alternative fuels, Brin, Larry Page, and his brother, Carl Page Jr., are investors in Nanosolar, Inc., a privately held California company that is developing solar cells for commercial, residential, and utility use. Nanosolar specializes in "thin-film solar cells"; the advantage of these cells is that they can be printed on plastic sheets that can be integrated into roofs, walls, and other surfaces transparently, eliminating the typical solar cell eyesore. The company has a $10.5 million grant from the federal Defense Advanced Research Projects Agency (DARPA), which provided funding for the creation of the Internet.

Given that one of Google's potential limitations on growth is the availability and cost of electricity, the involvement of Brin and Page in Nanosolar and in other energy-related experiments and investments is a logical extension of their future plans for Google and Google.org. The pair also plan to fund wireless Web access in various locations around the world. CEO Eric Schmidt sees his company's reach ultimately extending to every place on Earth. "When you look at the Amazon and you say, 'Why aren't there any Internet users?' it's because there is no power," he explains. "And people are working on this. So we'll get them all, even the people in the trees. It's just a matter of getting them power and some kind of a device."

While Brin, Page, and Google search for new sources of energy, the U.S. Department of Energy is investing heavily in genetics and biotechnology. The DOE is backing Craig Venter and related scientific research at an $80 million annual rate to support its own foray into genomics. The department's career employees are keenly aware of the role Google may play in contributing to solutions to some of the earth's biggest and most complex challenges.

Ari Patrinos, the point man for these DOE efforts, is a big fan and heavy user of Google. He turns to it an estimated 50 to 100 times daily, and fully appreciates its potential as a force and a partner in the search for answers to the world's dearth of clean, renewable sources of power. Both DOE and Google, each in its own way, are supporting biological research by Venter and others aimed at solving serious long-term problems.

"Google is getting into the biology business as they have gotten into other fields. I don't think the government has tried to do anything comparable to Google," Patrinos said. "We have been stressing the importance of advanced scientific computing research and information research that Google is helping to enable. It was an alien concept for most biologists until recently. The genome revolution has exploded in the production of vast amounts of data we need to analyze, process, and use. Search engines are extremely important for the biological data we have collected. It is the only way we will be able to exploit this treasure trove. The search engines have become sophisticated enough to identify functional elements of individual genes and proteins. These are not blind searches. There are pieces of this software that are almost like artificial intelligence."

Patrinos happens to live in Rockville, Maryland, next to Dr. Francis Collins, the director of NIH's National Human Genome Research Institute. He also maintains a close working relationship with Venter, and is trusted by all of the various parties. If NIH's primary focus is human health, the emphasis for Patrinos is analyzing the DNA of animals and plants to find fresh ways for bioremediation, the cleaning up of toxic sites, and the development of clean-burning fuels. It was Patrinos, in his basement over beer and pizza, who negotiated the compromise in 2000 that led to the joint announcement at the White House about the mapping of the human genome. Born in Athens, Greece, he earned a Ph.D. in mechanical engineering and the astronomical sciences at Northwestern University and holds the formal title of director of the Office of Biological and Environmental Research at the DOE. He is passionate about the fusing of technology and biological research to provide new age answers through Google.

Patrinos says Google has the capacity to provide insights into the function of genes; given the enormous complexity, it is essential to have sufficient computing power to model all of the different operations within the cell. "The wizardry of search engines is that they can race through databases and show relationships and bring to light answers. This is information that can be used by enterprising technology firms or this growing field of industrial and environmental biotechnology. Google is working on their version of it."

And then there are the personal experiences that have made Patrinos a major Google enthusiast. "I have found colleagues I hadn't seen in 40 years through Google. One guy is in Nigeria, one guy is in France and another guy is in Australia. I wouldn't have found them any other way."

Google has other friends in Washington too, and it will need them as it grows larger and branches out. Power and size breed mistrust, so inevitably there are calls from competitors and others for regulation and restrictions. Google has already encountered hostile opposition from some who fear it is trampling on their rights. After publishers raised legal objections, it chose to put a temporary moratorium on its scanning of copyrighted library books. Moreover, given the company's emphasis on gathering information it saves about millions of searchers, as well as its forays into genetics and biology, these far-flung interests will raise concerns about ethics and privacy that have strong political overtones. Sharon Terry, president of the Washington-based Genetic Alliance, is someone Google is likely to have on its side as it does battle in the political trenches.

Terry's journey into genetics began as a parent, when in 1994 her two children were diagnosed with a rare condition that leads to premature aging. She was seeking both authoritative information and other parents who were in the same situation. When Google came online, she used it to mine newsgroups and online bulletin boards. "It exploded how we could connect with other people," she said, adding that her kids are now 16 and 17 years old.

Terry, who holds an advanced degree in religious studies but has no training as a scientist, relies on Google daily to stay abreast of news and information about the disease and related genetic research. Has a group in Hungary posted a new paper on the disease? If so, she finds it and reads it immediately, thanks to Google. "I can get specialized information quickly by using Google," Terry says. "It allows the layperson to come up to speed on specific topics and get into the discussion in ways that I wouldn't otherwise be able to."

In her professional life leading the Genetic Alliance, Terry's goal is to make it the leading clearinghouse for people seeking information about genetics. She is deliberately nonpolitical, positioning the organization in a way that enables her to converse with researchers, policymakers, and private-sector pharmaceutical firms. The Alliance receives frequent questions from individuals looking for information, and Terry and her staff use Google and Google Scholar throughout the workday to help find answers. She remembers how much harder it was a decade ago, when search engines lacked Google's comprehensiveness and ability to rank search results by relevance.

"Nothing else meets the depth of Google to hone in on exactly the right information," Terry said. "Our work is connecting the dots, and Google connects the dots wonderfully. We may meet someone who says 'I have a concern about a certain condition,' and often we do a Google search first. I remember ten years ago when sites didn't have good indexes, and you had to wade through crap to get to the nut you needed. We have grown our organization and ability to do what we need to do exponentially, in a way that is attributable to the Internet and good search engines like Google. I can't imagine Google not existing."

Brin and Page want to make it even easier for Terry to find the information she needs. Their ambitions and at times seemingly crazy ideas go far beyond Google alone. People around the world may see Google and the Internet as one these days, but Brin and Page foresee the potential for human beings and the search engine to grow ever closer.

"Why not improve the brain?" Brin asked. "You would want a lot of compute power. Perhaps in the future, we can attach a little version of Google that you just plug into your brain. We'll have to develop stylish versions, but then you'd have all of the world's knowledge immediately available, which is pretty exciting."

APPENDIX I

23 Google Search Tips

1. Google can be your phone book. Type a person's name, city, and state directly into the search box, and Google will deliver phone and address listings at the top of the results. The feature works for business listings too.

Bonus tip: Google can also work as a reverse directory; if all you have is a phone number, type it in and any matches will appear in the results.

2. Google can be your calculator. Type a math problem into the search box and Google will compute it. You can spell out the equation in words (two plus two, twelve divided by three), use numbers and symbols (2+2, 12/3), or type in a combination of both (ten million * pi, 15% of six).

3. Longer is better, but shorter is okay. Google is designed to return high-quality results even for one- or two-word queries, so you can keep your searches short. But adding a few more words often yields better results.

Example: While gathering information on applying to colleges, include the word *admissions* after the name of a university you are searching to get more relevant results.

4. Use quotation marks when precision matters. Typing *"the search is over"* into Google will return Web pages about the rock song by Survivor—but leaving off the quotes will produce an assortment of unrelated pages. The reason: adding quote marks around a query tells Google to look for occurrences of the exact phrase as it was typed. That makes quote marks especially helpful when searching for song lyrics, people's names, or expressions such as *"to be or not to be"* that include very common words.

5. Google can be your dictionary. Type *define* followed by any English word into the search box, and Google will give you a quick definition at the top of the search results.

6. Capitalization doesn't matter. Save yourself time and typos: don't bother with the SHIFT key. Googling *Queen Elizabeth II* and *queen elizabeth ii* yields the same results. So whether you enter words in uppercase or lowercase, Google treats them equally—though the Queen would prefer otherwise.

Bonus tip: Google ignores common words like *the*, *and*, *is*, *of*, and *to* when they are used in search queries without quotes. Adding them will not change or improve your results, so you can leave them off.

7. Forget pluralism. Google automatically searches for all the stems of a word, so you don't need to do separate searches for *dance*, *dances*, and *dancing*. Just type one of the words and Google will take care of the rest, giving you results all in one list.

8. Get the picture. Looking for a photo of Paris Hilton, or the Paris Hilton Hotel? Click on the "Images" link above the search box, type your query, and Google will provide any photos or graphics in its database of over one billion images that match your terms, with a link to the page where they appear.

Bonus tip: Image searches may lead you to helpful Web sites that you otherwise might not find in the regular search results.

9. Maps, driving directions, and satellite views are one click away. The fastest way to finding the fastest way to your destination is to enter a city and state (or just a zip code) into the search box.

Example: Type *washington dc*, and Google will provide a direct link to its own map and directions service, maps.google.com, as well as those by Yahoo and MapQuest, making it easy to compare and get a second opinion. On Google Maps, you can toggle between a standard map view and overhead satellite imagery that pans and zooms with the drag of a mouse, and find listings for local businesses too.

10. Where do you want to go today? If you know the specific Web site you want to visit, type its name into the Google search box, hit enter, and you will be there in a flash.

11. Browse the world's bookshelves online. Search for a topic at print.google.com and you will see information from actual books that Google has scanned and indexed in its database. You can browse or read the entire text of works that are not copyrighted; for others, you can see snippets of pages where your search term appears and learn where to buy a full copy.

12. Dial GOOGL when you're on the go. Get phone numbers, directions, movie times, stock quotes, and more delivered to your cell phone. Send a text message with your query to the number 46645 (GOOGL on most U.S. phones) and the search engine will message you back with instant answers.

13. I'm Feeling Lucky. Enter a search term and click this button on the Google homepage to bypass a long list of results and go directly to the top-matching Web page for your term.

14. Google can be your newscaster. Google News, reachable via the "News" link above the search box or at news.google.com, provides up-to-the-minute information on politics, business, technology, entertainment, health, sports, and more. Type a topic of interest into the Google News search box to find the most recent stories from more than 4,500 global news sources.

Bonus tip: To follow a topic closely, sign up at google.com/alerts for regular alerts that are emailed to you with the most up-to-date news and Web links.

15. Google can be your weatherman. Type *weather* followed by a zip code or the name of a city, and Google will give the current conditions and a four-day forecast at the top of the results page.

16. Become a researcher. Google tends to list popular and fresh pages at the top of its results, but dig beyond the first page or two of search results and you will often find older, forgotten pages that have just what you need for a research project. Also check out the "Cached" versions of Web pages that Google collects as it crawls and downloads the Web, which are available through a blue link at the end of every search result. The cached version is an old version of the page, and often has the content you are seeking even if the current version of the page has changed—say, a news site that removed the original story.

Bonus tip: The cached version also highlights your search terms in color wherever they appear on the page, an especially helpful feature when combing through long documents.

17. Become a scholar. Serious searchers can tap into thousands of scientific and academic journals with Google Scholar. Enter a query into the search box at scholar.google.com to get abstracts and papers from published sources.

18. Take a magic ~ ride. The tilde character "~" in the corner of your keyboard is a handy tool in Google searches. Put it before a word, with no space between, to have Google look for pages with both that term *and* its synonyms.

Example: A search for *~auto* will also turn up Web pages that use the terms *cars, trucks, automobiles,* and more.

19. Pack more results onto each page. The "Preferences" link to the right of the search box is your ticket to tweaking various settings for Google searches, including the number of results displayed per page. Increase the number of matches you see per page from the standard set of 10 to 20, 30, or more, to put more answers at your fingertips faster.

20. Translate into other languages. The "Language Tools" link, also found to the right of the search box on the homepage, calls up Google's automated translation service as well as other language options. From this page, you can translate text among numerous languages (English to Spanish, French to German, Chinese to English...) or translate a Web page simply by entering its address.

21. Get an instant stock quote. Type a stock ticker symbol into the search box to get a stock quote and chart on any public company listed on the New York Stock Exchange, American Stock Exchange, or NASDAQ.

22. Get PG-rated results. A search on a serious topic like *sex education* might trigger objectionable material, so Google provides an optional SafeSearch filter to keep results family-friendly. Click the "Preferences" link next to the search box to view and adjust the SafeSearch settings (choose from "strict," "moderate," or no filtering).

23. Peer inside Google. Click the "more >" link above the search box to find additional Google features and products as well as further tips on how to search effectively. Check out the very handy one-page Google search guide at google.com/help/cheatsheet.html.

Bonus tip: See what the future of Google innovation holds—including TV search, personalized search, a real-time taxicab locator, and more—at Google Labs. Just type "Google Labs" into the Google search box. Goooooooood luck!

Appendix II

TEST CODE: WR-426F

GLAT®

GOOGLE LABS APTITUDE TEST®

How much aptitude do you have for the sort of mind-bending engineering problems encountered each day at Google Labs? Take the GLAT and find out. Simply answer all questions to the best of your abilities (cheaters will answer to the karma police), fold completed exam into attached envelope and send to Google Labs. Score high enough and we'll be in touch. Good luck.

Google(LABS)

PLEASE PRINT NEATLY. WE WILL NOT DISTRIBUTE OR DISCLOSE ANY OF YOUR PERSONAL INFORMATION. WE PROMISE. (IF YOU'D LIKE TO INCLUDE YOUR RESUME, WE'D LIKE TO SEE IT.)

Last Name | First Name

Email Address | Website

Daytime Phone | Evening Phone | Hair?

1. Solve this cryptic equation, realizing of course that values for M and E could be interchanged. No leading zeros are allowed.

   WWWDOT - GOOGLE = DOTCOM

   answer:

2. Write a haiku describing possible methods for predicting search traffic seasonality.

   answer:

3.
```
      1
     1 1
     2 1
    1 2 1 1
   1 1 1 2 2 1
```
   What is the next line?

   answer:

4. You are in a maze of twisty little passages, all alike. There is a dusty laptop here with a weak wireless connection. There are dull, lifeless gnomes strolling about. What dost thou do?

   - A) Wander aimlessly, bumping into obstacles until you are eaten by a grue.
   - B) Use the laptop as a digging device to tunnel to the next level.
   - C) Play MPoRPG until the battery dies along with your hopes.
   - D) Use the computer to map the nodes of the maze and discover an exit path.
   - E) Email your resume to Google, tell the lead gnome you quit and find yourself in whole different world.

5. What's broken with Unix? How would you fix it?

   answer:

6. On your first day at Google, you discover that your cubicle mate wrote the textbook you used as a primary resource in your first year of graduate school. Do you:

   - A) Fawn obsequiously and ask if you can have an autograph.
   - B) Sit perfectly still and use only soft keystrokes to avoid disturbing her concentration.
   - C) Leave her daily offerings of granola and English toffee from the food bins.
   - D) Quote your favorite formula from the textbook and explain how it's now your mantra.
   - E) Show her how example 17b could have been solved with 34 fewer lines of code.

1

7. Which of the following expresses Google's over-arching philosophy?

- A) "I'm feeling lucky"
- B) "Don't be evil"
- C) "Oh, I already fixed that"
- D) "You should never be more than 50 feet from food"
- E) All of the above

8. How many different ways can you color an icosahedron with one of three colors on each face?

answer:

What colors would you choose?

answer:

9. This space left intentionally blank. Please fill it with something that improves upon emptiness.

answer:

10. On an infinite, two-dimensional, rectangular lattice of 1-ohm resistors, what is the resistance between two nodes that are a knight's move away?

answer:

11. It's 2 PM on a sunny Sunday afternoon in the Bay Area. You're minutes from the Pacific Ocean, redwood forest hiking trails and world class cultural attractions. What do you do?

answer:

12. In your opinion, what is the most beautiful math equation ever derived?

answer:

13. Which of the following is NOT an actual interest group formed by Google employees?

- A. Women's basketball
- B. Buffy fans
- C. Cricketeers
- D. Nobel winners
- E. Wine club

14. What will be the next great improvement in search technology?

answer:

15. What is the optimal size of a project team, above which additional members do not contribute productivity equivalent to the percentage increase in the staff size?

- ○ A) 1
- ○ B) 3
- ○ C) 5
- ○ D) 11
- ○ E) 24

16. Given a triangle ABC, how would you use only a compass and straight edge to find a point P such that triangles ABP, ACP and BCP have equal perimeters? (Assume that ABC is constructed so that a solution does exist.)

answer:

17. Consider a function which, for a given whole number n, returns the number of ones required when writing out all numbers between 0 and n. For example, f(13)=6. Notice that f(1)=1. What is the next largest n such that f(n)=n?

answer:

18. What's the coolest hack you've ever written?

answer:

19. 'Tis known in refined company, that choosing K things out of N can be done in ways as many as choosing N minus K from N: I pick K, you the remaining.

Find though a cooler bijection, where you show a knack uncanny, of making your choices contain all K of mine. Oh, for pedantry: let K be no more than half N.

answer:

20. What number comes next in the sequence: 10, 9, 60, 90, 70, 66, ?

- ○ A) 96
- ○ B) 1000000000000000000000000000000000
0000000000000000000000000000000000
000000000000000000000000000000000
- ○ C) Either of the above
- ○ D) None of the above

21. In 29 words or fewer, describe what you would strive to accomplish if you worked at Google Labs.

answer:

Answers to the GLAT can be found at www.thegooglestory.com

APPENDIX III
Google's Financial Scorecard

In addition to being the world's most popular search engine, Google is also one of the world's most powerful financial engines. The company is a prolific cash machine, churning out billions of dollars in its short history. While many people follow Google's stock price closely, that is only one gauge of its performance. The amount of cash the business generates over time is actually a better measure of its underlying, long-term financial health. Since its founding in 1998, Google has produced billions of dollars in cash and has no borrowings. This has occurred while the company has reinvested substantial amounts of money to build the biggest and fastest computer network of its kind, one that handles more than 120,000 searches per minute, most in a fraction of a second.

Google's sales come from billions of clicks on Internet ads that appear alongside its search results, and on hundreds of thousands of other Web sites that are part of its vast network of affiliates. The company generates about half of its sales from Google.com and other Google-owned Web sites (Froogle, Gmail), and about half comes from Web sites of affiliates. As the company has grown larger, it also has become more prosperous. These days, more of each additional dollar of sales drops straight to the bottom line, fueling profits. (Financial and technology insiders refer to this trend by saying that Google's business is "highly scalable.") Keeping tabs on the company's rate of growth in the years ahead is likely to be one of the best and clearest ways to measure the increasing role of the Internet in people's lives.

While Google's prospects for growth appear limitless, that is folly. As magical as things seem during this honeymoon, the search engine is actually a real business run by mortals who will inevitably make mistakes, despite their strong track record and continuing desire to achieve superior results. Sergey Brin, Larry Page, and Eric Schmidt also anticipate that competition and other factors will slow its torrid rate of growth. Neither they nor anybody else can predict how and when that slowdown will occur. In the meantime, Google's hypergrowth is constrained by how rapidly it can hire and integrate new employees—Nooglers—in the U.S. and around the world. Someday, its prospects may be affected by the "unknown unknown"—the risk posed by unpredictable events in an uncertain future.

What follows are figures that help to illuminate Google's role as a robust financial enterprise.

THE GOOGLE FINANCIAL SCORECARD

$ in millions, except as noted

YEAR ENDED DECEMBER 31

	1999	2000	2001	2002	2003	2004
SALES	$220 thsnd	19.1	86.4	439.5	1.5 billion	3.2 billion
EXPENSES *(excluding income taxes)*	$6.7	33.8	75.5	253	1.1 billion	2.5 billion
PROFITS	−6.1	−14.7	7	99.7	105.6	399
CASH FROM OPERATIONS *(excluding investment in computing)*	–	–	31.1	155.3	395.4	977

On August 19, 2004, Google went public at a price of $85 per share. Trading under the stock ticker symbol GOOG, the shares more than tripled in price during its first year, trading as high as $317.80 on July 21, 2005. It never traded below the initial public offering price of $85.

On August 18, 2005, one year after the IPO, Google stock closed at $280. At this price, founders Sergey Brin and Larry Page each owned about $10 billion worth of Google shares. To diversify their investments, each of them sold about $750 million of stock during the firm's first year as a public company.

In the first half of 2005, Google's financial performance sizzled. The company's sales doubled, its profits shot up 500 percent, and its cash from operations grew 300 percent. By June 30, 2005, the company had nearly $3 billion in cash and no borrowings.

SIX MONTHS ENDED JUNE 30

	2004	2005
SALES	$1.4 billion	$2.6 billion
EXPENSES *(excluding income taxes)*	$1 billion	$1.7 billion
PROFITS	$143 million	$712 million
CASH FROM OPERATIONS *(excluding investment in computing)*	$370.6 million	$1.2 billion

Source: Google Inc.

On August 18, 2005, Google announced plans to raise about $4 billion by selling 14,159,265 additional shares of stock to investors. It was the largest secondary stock offering ever by a one-year-old media and technology firm. The unusual number of shares was meaningful to Sergey Brin and Larry Page, who are both mathematicians. The figure represented the first eight digits following the decimal point in the value of pi (3.14159265), a famous term from geometry derived by dividing the circumference of a circle by its diameter. The Google Guys continued to maintain voting control over all major decisions affecting the company after the $4 billion stock offering.

STOCK MARKET VALUE OF GOOGLE VS. OTHER U.S. CORPORATIONS AUGUST 18, 2005

Microsoft	$287 billion
Wal-Mart	$197 billion
Time Warner	$85.9 billion
Google	**$79.6 billion**
eBay	$54.8 billion
Disney	$52.8 billion
Yahoo	$48.5 billion
McDonald's	$42.3 billion
GM	$19.3 billion
Amazon	$18.2 billion

A Note on Sources

This book is based primarily on interviews with more than 150 people as well as video and audio recordings and thousands of pages of public and private documents, Internet postings, and emails. We interviewed dozens of current and former employees of Google, including officers and directors of the company, product managers, engineers, research and marketing specialists, and others. We spoke with numerous investors, clients, competitors, and partners of Google's, both past and present, including officials from AOL, Yahoo, Microsoft, AltaVista, and Ask Jeeves. We interviewed professors and administrators at Stanford University who know Larry Page, Sergey Brin, and other Google employees, and who advised and worked with them when they were students. We also talked with officemates, classmates, and friends of Larry and Sergey from Stanford. In addition, we spoke with dozens of search industry analysts and technology experts; Wall Street financial analysts; friends, family, and former classmates of Google employees; officials at the National Institutes of Health and other leading geneticists and scientists; government officials from the Department of Energy and other agencies; consultants and associates of Google's; library directors and officials at Stanford, Harvard, Michigan, Oxford, and the New York Public Library; Internet privacy and security watchdog groups; Google searchers both in the U.S. and abroad; and others.

Many of the interviews were conducted "on the record," and those individuals are named in the text wherever possible. Numerous other interviews were conducted "on background," meaning that we used the information these sources provided but agreed not to identify them by name. In many such instances, sources cited the need to remain anonymous to protect ongoing business or personal relationships with people at Google.

We wish to thank everyone, named and unnamed, who spoke with us and generously shared their time and insight. It is because of their guidance, patience, and trust that we were able to uncover and recount so many of the fascinating and compelling aspects of this company and its founders. We hope that we have fairly and accurately captured their words, experiences, and emotions in the narrative.

Much of the reporting for this book was done from the Washington DC area, where the tentacles of *The Google Story* extended in some significant and unexpected ways. We also spent considerable time in northern California and Silicon Valley, visiting the Googleplex, Stanford University, and other sites of interest. In addition, we reported and witnessed the global reach of Google during travels to Europe, South America, Israel, West Africa, and cities and towns across the United States. On the Internet, where much of this story takes place, we made use of archives stored on Google and the Wayback Machine at web.archive.org to view how Web sites appeared at specific moments in time.

The background for this book evolved over a period of years from beat reporting by David Vise on the intersection of technology, business, and the media for *The Washington Post*.

We are aware of the difficulties in any attempt to reconstruct events after the fact. Wherever possible, we reviewed the accounts of conversations and events with multiple persons involved or with other knowledgeable parties, to ensure that we had the best obtainable version of what happened. We also drew heavily on contemporaneous documents and materials that, unlike memories, do not change over time. In addition to video and audio recordings, these included filings with the Securities & Exchange Commission, court papers, transcripts, contemporaneous notes, archives of Internet pages, emails, slide presentations, and other documents. We attended, or obtained video footage of, numerous talks and presentations given by Google employees and others, including Sergey Brin and Larry Page's appearance at an Israeli high school and their explanation of Google's inner workings to an audience at Stanford. Additionally, we relied on videos of talks given by Eric Schmidt and Jeffrey Dean at the University of Washington. We also made use of public records, dated postings on Google's Web site, and surveys and studies of Internet usage compiled by Family Safe Media, the Pew Internet & American Life Project, and others. We also reviewed more than 100 pages of correspondence between the SEC and Google concerning the company's unorthodox initial public offering, obtained under a Freedom of Information Act request.

Google has received extraordinary attention from the media in its seven years as a company, and we were assisted enormously in our research by the high-caliber work of other journalists and authors who have covered the company. In some cases, we included quotations that first appeared in articles written by others, particularly when those stories appeared in a Q&A format or transcript. This provided the considerable benefit of seeing the statements, thoughts, and opinions of Brin, Page, and other key players at specific moments in the evolution of Google, rather than just looking back after the fact. Where it did not break the story's narrative, we identified the sources of those quotations in the text. Many of them are also credited here.

We drew heavily on John Heilemann's excellent reporting of Google's IPO day and financial history in his article, "Journey to the (Revolutionary, Evil-Hating, Cash-Crazy, and Possibly Self-Destructive) Center of Google," which ran in *GQ* magazine in April 2005. We also relied on the August 2004 *Playboy* interview of Sergey Brin and Larry Page that was reprinted in a public SEC filing by the company. Two long, thoughtful articles on Google's business prospects and battle against Microsoft that were particularly helpful were Fred Vogelstein's "Search and Destroy," in *Fortune* (May 2, 2005) and Charles H. Ferguson's "What's Next for Google," in the *MIT Technology Review* (January 2005).

Articles by Adam Tanner of Reuters and Tom Howell of the University of Maryland *Diamondback* provided information about Sergey Brin's childhood and family that we otherwise would not have known. Reporting by Scott Anderson for the *Ann Arbor News* and a profile in the *Michigan Engineer* informed our account of Larry Page's early years. Michael Specter's article "Search and Deploy" from the May 29, 2000, issue of *The New Yorker* offered a fascinating slice-in-time look at Google before it was a juggernaut. In relating the experience of Burning Man, we relied in part on Brian Doherty's book *This Is Burning Man,* as well as vivid reporting in *Reason* magazine, the *San Francisco Chronicle*, Wired News, and the festival's own *Black Rock Gazette.* We made use of *Wired* magazine's reporting on how the rich and famous use Google, which appeared in the March 2004 article "Googlemaniacs." Charlie Rose conducted a number of thought-provoking interviews with Eric Schmidt and other technologists for his PBS show, which proved invaluable to us. Also helpful was Lesley Stahl's segment on Google for the CBS show *60 Minutes* in January 2005. Blogger Brad DeLong's notes from a Larry Page speech in 2003 yielded an amusing nugget about Google's first profits.

To follow the daily news coming out of Google and the search and technology industries, we relied on excellent coverage in *The Washington Post*, *The Wall Street Journal*, *The New York Times*, the *San Jose Mercury News*, and *BusinessWeek.* The Web versions of the above publications, as well as The Associated Press, Reuters, Cnet.com, Marketwatch.com, and Wired News provided top-quality reporting and analysis—in many cases almost instantaneously. We were especially grateful to have a resource like Danny Sullivan's SearchEngineWatch.com at our disposal while writing this book. We learned a great deal from Danny's newsletters, Web site, and Search Engine Strategies conferences.

We also benefited from the business and technology coverage on CNBC, NPR, and the BBC, and in the *San Francisco Chronicle*,

The Sun (Baltimore), *The Boston Globe, The Seattle Times, Silicon Valley Business Journal, USA Today, Barron's, Fast Company, Forbes, Fortune, Newsweek, Time*, and *Wired*. Publications based outside the U.S., including *The Economist, Financial Times,* and *The Independent* (U.K.), *The People's Daily* (China), and *Le Monde* (France), aided our narrative. We relied on the helpful news aggregation provided by Google News, Yahoo News, Slashdot, and MediaPost. Countless personal Web sites and blogs that echo and comment on search industry happenings also added to our understanding and provided a safety net for catching news nuggets that otherwise might have eluded us.

Several books influenced how we thought about and framed *The Google Story*, including Tracy Kidder's classic *The Soul of a New Machine*; Katie Hafner and Matthew Lyon's account of the birth of the Internet, *Where Wizards Stay Up Late*; the eBay story, compellingly told by Adam Cohen in *The Perfect Store;* Karen Angel's *Inside Yahoo!: Reinvention and the Road Ahead*; and John Markoff's *What the Dormouse Said*.

If we have inadvertently failed to cite any published works or other sources from which we have benefited in telling *The Google Story*, we apologize in advance.

Our first and last resource in writing this book was Google itself. We turned to the search engine many times each day for news, research, images, email, maps, and corporate information. In these ways and more, this book would not have been possible without Google.

Acknowledgments

This book would not have been possible without the enormous support of *The Washington Post,* a newspaper with a soul. There is no better place to be a journalist. Our thanks to Post Company chairman and CEO Don Graham, publisher Bo Jones, executive editor Leonard Downie Jr., managing editor Phil Bennett, business editor Jill Dutt, photo editor Joe Elbert, and Dan Beyers, a terrific hands-on editor. Thanks also to research maven Richard Drezen for numerous assists.

Our thanks to Google co-founders Sergey Brin and Larry Page, and CEO Eric Schmidt, for trusting us to tell their story. While we maintained total editorial control over the book's contents, we discussed the manuscript with Google executives prior to publication to ensure maximum fairness and accuracy. Special thanks to Google corporate P.R. chief David Krane and spokesman Steve Langdon for providing insight, arranging interviews with Googlers in Silicon Valley and around the world, and assisting us with Google doodles and photographs. During this demanding period of rapid growth, we greatly appreciate your making time for *The Google Story*.

We particularly want to express our gratitude to literary agent Ron Goldfarb. His initial enthusiasm for *The Google Story* made the difference in finding the right publisher and editor for this book, and his instincts at every turn have helped us immensely. Thanks to Nita Taublib, deputy publisher of the Bantam Dell division of Random House, who acquired the book and supported the idea from the start, and to Irwyn Applebaum, Bantam Dell's president and publisher. Special thanks to Ann Harris, executive editor of Bantam Dell, whose editing and questions improved the book markedly. Thanks also to Ann's editorial assistant Meghan Keenan, art director Glen Edelstein, title administrator Loren Noveck, and copy editor Peggy McPartland. Under deadline pressure, all of you made this a priority without compromising on quality. This book is being published in many languages around the world, and our thanks to Sharon Swados at Random House for handling the foreign rights.

We are also very grateful to the many individuals who offered us ideas, contacts, and support along the way. We would especially like to thank Joe Perpich for his friendship and intellectual engagement with this project. A renaissance man, Joe was the first to encourage us to explore the significance of the intersection of genetics, scientific research, and Google. He opened doors that made it possible for us to take readers to a new level of understanding about Google.

From David:

First and foremost, thanks to my wife Lori for encouraging me to tackle *The Google Story* and other worthy challenges; for listening, reading the manuscript, and offering excellent suggestions along the way; and for striving for excellence in everything you do and encouraging me to do the same. Your sacrifices, generous spirit, and love enabled me to throw myself fully into the reporting and writing of this book. I could not have done it without you, and I love you very much.

Special thanks to my remarkable parents, Doris and Harry Vise, for their wise counsel and love in connection with this book and so much else since the day I was born. Actually, now that I think about it, I was born in the middle of the night, which must be why that is my favorite time to write.

Many thanks to: our terrific daughters, Lisa, Allison, and Jennifer, who have mastered the art of instant messaging, talking on the phone, and Googling simultaneously; my sisters, Judy Schaengold and Joyce Vise; my mother-in-law, Toby Silverman; our nieces, Rachel and Jessica Schaengold; my sisters-in-law, Jan Allen and Jamie Be; and my brothers-in-law, Mark Schaengold and Bobby Silverman. Each of you contributed meaningfully to *The Google Story.*

From Mark:

Thanks to Bruce Brumberg, Howard Zaharoff, and Matt Obernauer for guidance and encouragement at the start of this project, and to Alex Robbins for that first push years ago. Special thanks to Bob Woodward and Elsa Walsh for their trust, generosity, sage advice and friendship. I also owe a debt of gratitude to Katie Kessenich for cheerfully letting me turn our apartment into an all-hours office, and to Lori and the girls for always welcoming me into their home. Thanks also to Caryn Aulenbach for her hospitality on trips to the West Coast.

For always being there, Jenny, whether you were nearby or far away, you have my deepest appreciation and affection. I look forward to many more adventures and travels together.

I owe much to my small but wonderful family. Thanks to my sister, Natalie, for always challenging and entertaining me, and for being a great friend. To my grandmother for her love, support, and nourishment. And especially to my parents, Roger and Zoriana, for a lifetime of loving, caring, and teaching. You are still the first place I turn for the things that matter most.

Photo Credits

Larry and Sergey in a Menlo Park garage: © william mercer mcLeod.

Early Google computer partially made from imitation LEGO blocks: Courtesy of Google.

John Doerr: © 1998, *The Washington Post*. Photo by Todd Cross. Reprinted with permission.

Michael Moritz: Courtesy of Sequoia Capital.

Craig Silverstein: Courtesy of Google.

Marissa Mayer: Courtesy of Google.

Early Google computer server: Till Moepert.

Charlie Ayers: Courtesy of Charlie Ayers.

Danny Sullivan and Sergey Brin at a "fireside chat": Judy Miller.

Sergey with Larry on scooter at the Googleplex: Kim Kulish Photography.

Googleplex campus as seen from a B-24 bomber: Andy Meyer.

Google Internet café in Cairo: Daniel Doubrovkine.

Matt Cutts, aka "the Porn Cookie Guy": AP Photo/John Todd.

Sergey and Larry in their office hot tub: © william mercer mcLeod.

Craig Venter: Courtesy of J. Craig Venter Institute.

Playboy interview: Courtesy of *Playboy*; photographs © 2003 Kim Kulish.

Larry presiding over NASDAQ stock exchange opening: © 2004, The Nasdaq Stock Market, Inc.

Larry, Sergey, and CEO Eric Schmidt: Markham Johnson Photography.

Index